politicians and the public at large to understand not only the overwhelming scientific evidence that has emerged in recent years, but also the remaining uncertainties that need to be addressed in future scientific endeavor. This feature alone and the simple and readable manner in which the book is written make it essential reading for scientists as well as the concerned public at large."

Dr. R.K. Pachauri, Chairman, Intergovernmental Panel on Climate Change (IPCC) and Director General, The Energy and Resources Institute (TERI), India

"... there is a real need for a comprehensive book on climate change ... *The Science and Politics of Global Climate Change* is it. It does exactly what the title and subtitle promise, providing insights into the causes and effects of the contributing meteorological phenomena and into why it has been so hard to get consensus among governments ... copies should be shipped to anyone who doubts the reality of climate change, starting with presidents in denial."

New Scientist

"... requires no specialised knowledge, but is accessible to any educated general reader who wants to make more sense of the climate change debate. It also sheds light on how science is used in policy debates."

The Chemical Engineer

"Each of the key aspects of global climate change is covered, with up-to-date and well-referenced information throughout. Its impressive breadth and the provision of succinct overviews of source material in the further reading sections of each chapter mean that teachers, lecturers and researchers will all find this book a useful starting point for in-depth study. There are now numerous taught masters courses on 'global change issues', and this book constitutes a must-have addition to their reading lists ... read the book in its entirety – it is well worth it. ...

"[This book] is an excellent attempt at deconstructing the confusion that surrounds the climate-change debate. This reviewer has been waiting some time for a book such as this to appear ... The science and politics of climate change are brought together quite seamlessly ... Dessler and Parson's book is a must for those who want to move beyond the rhetoric and understand the relationship between climate science and policy, and also for those seeking an interdisciplinary outlook on the management of global environmental issues. ...

"This book will be most useful to undergraduates and post-graduates in the fields of environmental science, sustainability and international politics ... as a primer that brings together global climate change science and politics it succeeds very well indeed."

Times Higher Education Supplement

"This is an excellent way into the subject for the beginner ... one of the most lucid and readable introductory accounts of the topic that has been published in some while. As such it should be seen as a 'must-buy' and an essential addition to the library."

TENews

"This is a book which all scientists and the educated general public should read and reflect upon before it is too late to halt the apparently inevitable progress to Armageddon."

Chromatographia

"... a useful compendium of the current debates in the science and politics of climate change ... succinct and consistent book ... Ensure[s] fluent reading for non-expert, yet educated, citizens. The book is logically structured and it should become a key reading and teaching source in geography and environmental sciences. It can also be valuable to doctoral students and senior researchers interested in learning about climate change science and politics. Overall it is a book worth having on one's shelf."

Environmental Sciences

"As more and more extreme weather events around the globe are being associated with climate change, it is sometimes difficult to be able to see the wood for the trees, but this book takes the reader very clearly through the 'maze' of claims and counter-claims. ... if only government leaders would read, digest and follow up some of the suggestions in the last chapter, there would be optimism that the problem can be overcome. As always with Cambridge University Press, the book, which is illustrated with diagrams, charts and boxes, is impeccably produced, and is an absolute 'must' for every reader of this journal."

International Journal of Meteorology

"Written by an atmospheric scientist and a law professor with extensive public policy experience, the book effectively tackles the rough-and-tumble intersection of science and policy that has led to confusion and inaction ... The scholarly value of [the book] is indisputable. Dessler and Parson independently possess significant authority on both the science and the politics of climate change. Their treatment of the subject illustrates the complexity of the problem with remarkable ease and clarity ... the carefully thought-through recommendations make this book critical reading for policymakers ... considering action on the issue."

Maria Ivanova, The College of William and Mary

betwee... ...t n...
contentious a... ...is new e...
to reflect the rapid mo... ...nt of events r...
the recent findings of the Intergovernmenta...
the Stern Review on the economics of climate c...
initiatives up to the 2009 Copenhagen meeting. In ...
the book have been improved; in particular a more th...
basic science of climate change is included. The book als...
the discussion of contrarian claims with the discussion of cu...
knowledge, extends the discussion of cost and benefit estimate...
an improved glossary.

ANDREW DESSLER is a professor in the Department of Atmospheric Scien...
at Texas A&M University. He received his Ph.D. in Chemistry from Harvard
in 1994. He did postdoctoral work at NASA's Goddard Space Flight Center
(1994–1996) and then spent nine years on the research faculty of the University
of Maryland (1996–2005). In 2000, he worked as a Senior Policy Analyst in the
White House Office of Science and Technology Policy, where he collaborated
with Ted Parson. Dessler's academic publications include one other book: *The Chemistry and Physics of Stratospheric Ozone* (Academic Press, 2000). He has also
published extensively in the scientific literature on stratospheric ozone
depletion and the physics of climate.

EDWARD PARSON is Joseph L. Sax Collegiate Professor of Law and Professor
of Natural Resources and Environment at the University of Michigan. His

research examines in
technology in public polic
book, *Protecting the Ozone Layer:*
2003), won the 2004 Harold and Marg
Studies Association. His academic articles
Climatic Change, *Issues in Science and Technology*, th
and the *Annual Review of Energy and the Environment*. Pa
on several senior advisory committees for the US Nationa
and US Global Change Research Program, and has worked and
for the International Institute for Applied Systems Analysis, the Un
Nations Environment Program, the Office of Technology Assessment of t
US Congress, the Privy Council Office of the Government of Canada, and the White House Office of Science and Technology Policy, where he collaborated with Andrew Dessler. In 2005, he was appointed to the National Advisory Board of the Union of Concerned Scientists. Parson spent 12 years on the faculty of Harvard's Kennedy School of Government. He holds degrees in Physics from the University of Toronto and in Management Science from the University of British Columbia, and a Ph.D. in Public Policy from Harvard.

Praise for the First Edition

"This timely, informative and well-written book does an excellent job of explaining, in language accessible to everyone, the scientific basis for our current understanding of global warming and climate change, as well as societal implications and the political barriers to sound, rational policy. Its co-authors are well recognized experts in science and in public policy. I recommend it to anyone who wishes to gain a better understanding of this complex issue – what the debate is all about – and as a core textbook for introductory courses on the environment, climate change, or public policy"

Professor Neal Lane, Malcolm Gillis University Professor and Senior Fellow of the James A. Baker III Institute for Public Policy, Rice University. Former Science Advisor to President Clinton and former Director of the US National Science Foundation

"As the scientific evidence on human induced climate change becomes stronger and more widely accepted, voices that question it appear to get louder and seemingly more coordinated. In a complex area such as climate change, politics inevitably runs into conflict with the domain of science. This book is a timely analysis of the scientific evidence of climate change as well as the political forces that question its full acceptance. Dessler and Parson have produced a remarkable piece of work that is relevant for the scientific community in understanding the political implications of their work and for

"… coverage and presentation of climate science and policy [is] commendable … a good candidate for a primer for multidisciplinary classes devoted to climate policy"

Randall M Wigle, Wilfrid Laurier University

"… both insightful and engaging … the book is also highly readable and well suited to reach a wide audience. That's good, because the gaps in understanding between scientists, policy makers, journalists, and the public remain a major barrier to the adoption of sensible responses to the problem. Dessler and Parson's book will help because it provides us with a sound and thoughtful guide to the climate change debate … It explains scientific and policy debates, discusses areas of knowledge and uncertainty regarding climate change, and offers possible policy options."

American Meteorological Society

"[Dessler and Parson] open with a powerful organizing principle for the climate and their book: to clearly distinguish between objective understanding (i.e. what we know) and subjective value judgment (i.e. what we believe should be). As a framework for thinking, this holds great promise: it curbs the potential to use ignorance to manipulate the debate, but also acknowledges the limits of scientific understanding."

Paul A. T. Higgins, Senior Fellow, American Meteorological Society Policy Program

"This important book will be as valuable to informed lay persons as to policy makers in government, industry and the wider community. While this is a political book, it is also a valid general primer about the science-policy interface. Furthermore, it serves as a model of how to clarify complex problems for lay persons through careful logic, wise choice of supporting literature, consistent definition of technical terms and clear writing."

Ecoscience

"… Dessler and Parson succeed in making both science and policy accessible to a wide readership. As someone working at the interface of science and policy, I could comfortably recommend this book to friends and colleagues. The book – which is well illustrated with easy-to-grasp figures, and which has summary tables provided at several key junctures – would also make an excellent resource for a high school or college-level survey course in either environmental studies or public policy."

Wendy S. Gordon, EOS

"Each of the key aspects of global climate change is covered, with up-to-date and well-referenced information throughout. Its impressive breadth and the provision of succinct overviews of source material in the further reading sections of each chapter mean that teachers, lecturers and researchers will all find this book a useful starting point for in-depth study."

David Reay, Edinburgh University

"Provides perhaps the most comprehensive and comprehensible analysis of the debates around climate change and is likely to become a foundational text for students, scholars, policymakers, and citizens seeking clarity on this topic. The scholarly value of *The Science and Politics of Global Climate Change* is indisputable. Dessler and Parson independently possess significant authority on both the science and the politics of climate change. Their treatment of the subject illustrates the complexity of the problem with remarkable ease and clarity. By juxtaposing the scientific and the political processes, they enrich the academic literature which has traditionally separated the two and open up new avenues for exploring policy solutions. Scientists will find value in the discussion of how their work is used by policymakers. Those knowledgeable about the politics of climate change will find value in the discussion of the science."

Global Environmental Politics

"Excellent overview of an increasingly critical issue."

Future Survey

"I found the book quite well written, with a good explanation of a suitable range of relevant scientific, 'political' and economic concepts ... I believe it is a good candidate for a primer for multidisciplinary classes devoted to climate policy"

Canadian Public Policy, Randall M. Wigle, Wilfrid Laurier University

The Science and Politics of Global Climate Change

A Guide to the Debate

SECOND EDITION

ANDREW E. DESSLER
Department of Atmospheric Sciences,
Texas A&M University

EDWARD A. PARSON
Law School and School of Natural
Resources and Environment,
University of Michigan

CAMBRIDGE UNIVERSITY PRESS
Cambridge, New York, Melbourne, Madrid, Cape Town, Singapore,
São Paulo, Delhi, Dubai, Tokyo

Cambridge University Press
The Edinburgh Building, Cambridge CB2 8RU, UK

www.cambridge.org
Information on this title: www.cambridge.org/9780521737401

First published 2006
Seventh printing 2009
Second edition 2010

Printed in the United Kingdom at the University Press, Cambridge

A catalog record for this publication is available from the British Library

Library of Congress Cataloging in Publication data
Dessler, Andrew Emory.
The science and politics of global climate change : a guide to the debate /
Andrew Dessler, Edward A. Parson. – 2nd ed.
p. cm.
Includes bibliographical references and index.
ISBN 978-0-521-73740-1 (pbk.)
1. Climatic changes. 2. Climatic changes–Government policy.
I. Parson, Edward. II. Title.
QC903.D47 2010
363.738'74–dc22
2009046537

ISBN 978-0-521-73740-1 Paperback

Additional resources for this publication at www.cambridge.org/dessler

Contents

Preface to the Second Edition

In the three years since the first edition of this book appeared, events related to climate change have moved rapidly. The fourth IPCC assessment report has documented the continued strengthening of scientific evidence for climate change, its predominant human causes, and the likely rate and risks of continuing changes. The Stern Review and subsequent debate have provoked a more serious discussion of the long-run character of climate-change risks and the appropriate way to evaluate them. The Kyoto Protocol first commitment period has arrived with many nations failing to meet their commitments, even as discussions starting in Bali in 2007 and continuing through Copenhagen in 2009 have sought to re-energize international actions. The United States has re-engaged with international efforts to build an effective response to climate change. And significant policy initiatives have been advanced by many nations, including comprehensive climate and energy legislation being considered in the US Congress. The accumulation of these events required this rewriting, even if continued rapid movement of climate-change policy and politics may mean that the summary of recent events in this edition will also have a short shelf-life.

But not everything about climate change moves fast. On the contrary, many key elements of the issue have changed little since we wrote the first edition. Although climate has many layers of variation on multiple time-scales, the basic dynamics of greenhouse-gas driven climate change operate on time-scales of decades and longer. Similarly in the energy system, the largest source of human-driven climate disruptions, the basic dynamics of capital turnover and technological change operate over decades. This is why – as we illustrate with the analogy of steering a supertanker that closes the book – most potential interventions to limit or respond to climate change only exert their full effect over decades of effort. Interventions to change course must be made well in advance of their effects, and in the face of considerable uncertainty. Scientific knowledge of climate change also advances slowly, for the most part. Many fields of research relevant to climate change are now fairly mature, so the major uncertainties about climate change, associated risks, and potential responses are increasingly well characterized. Sudden breakthroughs that would substantially change current understanding could happen, but appear rather unlikely.

It is relative to this fast-slow tension that recent events must be understood. The core structure of the climate problem can remain essentially unchanged, even while changes in public concern, prevailing framings, and political declarations and actions have transformed its surface. So the central conclusion of the first edition remains essentially unchanged: serious action on climate change must start immediately to avoid grave risks, and the urgency for action increases and the opportunities to avoid risks at low cost decrease with each year of delay. The flood of activity now occurring in multiple nations and internationally *might* add up to that first serious step needed to begin re-orienting investment toward the required transformations – but this is not yet clear. As we discuss within, while we share the widespread hope that the climate challenge is at last truly engaged, we have little confidence in the success of the current international process. And in case current efforts continue failing to deliver the required rapid, concrete progress, we propose an alternative approach that emphasizes strong unilateral leadership, coupled with action by a small group of major nations.

In addition to updating the discussion to reflect recent developments in the science, technology, economics, policy, and politics of climate change, we have also used the opportunity to strengthen a few parts of the first edition that we found weak. The most important change is to provide a more thorough primer on the basic science of atmospheric radiation that underlies climate change. In addition, we have integrated the discussion of contrarian claims with the discussion of current scientific knowledge in Chapter 3; extended the discussion of cost and benefit estimates, including a new section elaborating on the basis of current controversies in inter-temporal valuation and discounting; and provided an improved glossary.

Beyond these changes, the aims of the book and its intended audience are unchanged. Its key contribution remains bringing together a basic presentation of issues in science, technology, economics, policy, and politics as they pertain to climate change and highlighting the interactions among these domains, to provide a well founded understanding of where we are, how we got here, and where we need to go. With this breadth, it is targeted at the educated non-specialist reader seeking an introduction to the climate-change issue. In addition, for readers who are involved in climate issues from one side or another – the science, the policy, or the politics – the book aims to help them see how their piece fits into the bigger puzzle. In teaching, the book remains suitable for college courses at the upper-level undergraduate or introductory graduate/professional level, on climate change, environmental policy and politics, or science and public policy, with a prior course in physics, chemistry, or Earth science helpful but not necessary.

The few years since we worked on the first edition have deepened our own sense of the urgency and peril of addressing climate change. With this in mind, we dedicate our efforts in this new edition to our children: Matthew, Joshua, Alexander, and Michael. We do this in the hope that a prudent and competent global response to climate change – surely not too much to ask! – can preserve for them the opportunities for a secure, prosperous, and fulfilling life, and for connection with an undegraded natural world, that we have enjoyed.

College Station, Texas
Ann Arbor, Michigan

Preface to the First Edition

The Kyoto Protocol, the first international treaty to limit human contributions to global climate change, entered into force in February 2005. With this milestone, binding obligations to reduce the greenhouse-gas emissions that are contributing to global climate change came into effect for many of the world's industrial countries.

This event has also deepened pre-existing divisions among the world's nations that have been growing for nearly a decade. The most prominent division is between the majority of rich industrialized countries, led by the European Union and Japan, which have joined the Protocol, and the United States (joined only by Australia among the rich industrialized nations), which has rejected the Protocol as well as other proposals for near-term measures to limit greenhouse-gas emissions. Even among the nations that have joined Kyoto, there is great variation in the seriousness and timeliness of the emission-limiting measures they have adopted, and consequently in their likelihood of achieving the required reductions.

There is also a large division between the industrialized and the developing countries. The Kyoto Protocol only requires emission cuts by industrialized countries. Neither the Protocol nor the Framework Convention on Climate Change, an earlier treaty, provides any specific obligations for developing countries to limit their emissions. This has emerged as one of the sharpest points of controversy over the Protocol – a controversy that is particularly acute since the Protocol only controls industrialized-country emissions for the five-year period 2008–2012. In its present form, it includes no specific policies or obligations beyond 2012 for either industrialized or developing countries. While the Kyoto Protocol represents a modest first step toward a concrete response to climate change, there has been essentially no progress in negotiating the larger, longer-term changes that will be required to slow, stop, or reverse any human-induced climate changes that are occurring.

As these political divisions have grown sharper, public arguments concerning what we know about climate change have also grown more heated. Climate change may well be the most contentious environmental issue that we have yet seen. Follow the issue in the news or in policy debates and you will see arguments over whether or not the climate is changing, whether or not human activities are causing it to change, how much and how fast it is going to change in the future, how big and how serious the impacts will be, and what can be done – at what cost – to slow or stop it. These arguments are intense because the stakes are high. But what is puzzling, indeed troubling, about these arguments is that they include bitter public disagreements, between political figures and commentators and also between scientists, over points that would appear to be straightforward questions of scientific knowledge.

In this book, we try to clarify both the scientific and the policy arguments now being waged over climate change. We first consider the atmospheric-science issues that form the core of the climate-change science debate. We review present scientific knowledge and uncertainty about climate change and the way this knowledge is used in public and policy debate, and examine the interactions between political and scientific debate – in effect, to ask how can the climate-change debate be so contentious and so confusing, when so many of the participants say that they are basing their arguments on scientific knowledge.

We then broaden our focus, to consider the potential impacts of climate change, and the available responses – both in terms of technological options that might be developed or deployed, and in terms of policies that might be adopted. For these areas as for climate science, we review present knowledge and discuss its implications for action and how it is being used in public and policy debate. Finally, we pull these strands of scientific, technical, economic, and political argument together to present an outline of a path forward out of the present deadlock.

The book is aimed at an educated but non-specialist audience. A course or two in physics, chemistry, or Earth science might make you a little more comfortable with the exposition, but is not necessary. We assume no specific prior knowledge except the ability to read a graph. The book is suitable to support a detailed case-study of climate change in college courses on environmental policy or science and public policy. It should also be useful for scientists seeking to understand how science is used – and misused – in policy debates.

Many people have helped this project come to fruition. Helpful comments on the manuscript have been provided by David Ballon, Steve Porter, Mark Shahinian, and Scott Siff, as well as seminar participants at the University of British Columbia, the University of Michigan School of Public Health, and the University of Michigan Law School. A. E. D. received support for this project from a NASA New Investigator Program grant to the University of Maryland, as well as from the University of Maryland's Department of Meteorology and College of Computer, Mathematical, and Physical Sciences. All these contributions are gratefully acknowledged. A. E. D. especially notes the contributions of Professor David Dessler, for discussions in which many of the early ideas for the book were developed or refined.

1

Global climate change: a new type of environmental problem

1.1 The climate-change controversy

Of all the environmental issues that have emerged in the past few decades, global climate change is the most serious, and the most difficult to manage. It is the most serious because of the severity of harms it might bring. Many aspects of human society and well-being – where we live, how we build, how we move around, how we earn our livings, and what we do for recreation – still depend on a relatively benign and narrow range of climatic conditions, even though this dependence has been reduced and obscured in modern industrial societies by their wealth and technology. This dependence on climate can be seen in the economic harms and human suffering caused by the climate variations of the past century, such as the "El Niño" cycle and the multi-year droughts that occur in western North America every few decades. Climate changes projected this century are much larger than these twentieth-century variations, and their human impacts are likely to be correspondingly greater. Moreover, climate does not just affect people directly: it also affects all other environmental and ecological processes, including many whose connection to climate might not be immediately recognizable. Consequently, large or rapid climate change will represent an added threat to other environmental issues such as air and water quality, endangered ecosystems and biodiversity, and threats to coastal zones, wetlands, and the stratospheric ozone layer.

Projections of future climate change are uncertain, of course. Knowledge about climate change, like all scientific knowledge, is subject to uncertainty. We will discuss uncertainty, and how to make decisions about climate change under uncertainty, extensively in this book. But just because something is uncertain does not imply any particular advice on what to do about it. In particular, it does not necessarily mean the right course is to do nothing until we are certain. We do not wait to be certain the illness is life-threatening before calling the doctor, or wait to be certain we are going to drive into the tree before steering away from it. Sometimes we take action only when we are very confident it is the right course, but other times we take precautions against even rather unlikely risks. It depends on the particulars of each case.

For climate change, a key point about uncertainty is that it cuts both ways. Starting with some best estimate of climate change this century, making the estimate uncertain means that actual changes may turn out to be smaller than the current estimate, or bigger. Unless we prefer to run high-stakes risks – which people usually do not – this means uncertainty makes climate change more serious, not less. And the stakes are large. Present projections of climate change this century include, at the upper end of the uncertainty range, sustained rapid changes that appear to have few precedents in the history of the Earth, and whose impacts on human well-being and society could be catastrophic. This does not mean such extreme changes are certain, or even likely – but only that they are serious enough to be weighed in our decisions.

In addition to being the most serious environmental problem society has yet faced, climate change will also be the most difficult to manage. Environmental issues often carry difficult tradeoffs and political conflicts, because solving them requires limiting some economically productive activity or technology that is causing unintended environmental harm. Such changes are costly and generate opposition. But for previous environmental issues, technological advances and sensible policies have enabled large reductions in environmental harm at modest cost and disruption, so these tradeoffs and conflicts have turned out to be quite manageable. Controlling the sulfur emissions that contribute to acid rain in the United States provides an example. When coal containing high levels of sulfur is burned, in electric generating stations or other industrial facilities, sulfur dioxide (SO_2) in the smoke acidifies the rain that falls downwind of the smokestack, harming lakes, soils, and forests. Over the past 20 years, a combination of advances in technologies to remove sulfur from smokestack gases, and well-designed policies that give incentives to adopt these technologies, burn lower-sulfur coal, or switch to other fuels, have brought large reductions in sulfur emissions at a relatively small cost and with no disruption to electrical supply.

Climate change will be harder to address because the activities causing it – mainly burning fossil fuels for energy – are a more essential foundation of world economies, and are less amenable to simple technological correctives, than the causes of other environmental problems. Fossil fuels provide nearly 80 percent of world energy supply, and no alternatives now available could replace this huge energy source quickly or cheaply. Consequently, climate change carries higher stakes than other environmental issues, both in the severity of potential harms if the changes go unchecked, and in the apparent cost and difficulty of reducing the changes. In this sense, climate change is the first of a new generation of harder environmental problems that society will face this century, as the increasing scale of human activities puts pressure on ever more basic planetary-scale processes.

When policy issues have high stakes, it is typical for policy debates to be contentious. Because the potential risks of climate change are so serious, and the fossil fuels that contribute to it are so important to the world economy, we would expect to hear strong opposing views over what to do about climate change – and we do. But even given the issue's high stakes, the number and intensity of contradictory claims advanced about climate change is extreme. The following published statements give a sense of the range of views about climate change.

Former US Vice-President Al Gore:

> "So today, we dumped another 70 million tons of global-warming pollution into the thin shell of atmosphere surrounding our planet, as if it were an open sewer. And tomorrow, we will dump a slightly larger amount, with the cumulative concentrations now trapping more and more heat from the Sun. As a result, the Earth has a fever. And the fever is rising. The experts have told us it is not a passing affliction that will heal by itself. We asked for a second opinion. And a third. And a fourth. And the consistent conclusion, restated with increasing alarm, is that something basic is wrong. We are what is wrong, and we must make it right.
>
> "We, the human species, are confronting a planetary emergency – a threat to the survival of our civilization that is gathering ominous and destructive potential even as we gather here. But there is hopeful news as well: we have the ability to solve this crisis and avoid the worst – though not all – of its consequences, if we act boldly, decisively and quickly."[1]

United States Senator and former presidential candidate John McCain:

> "The burning of oil and other fossil fuels is contributing to the dangerous accumulation of greenhouse gases in the Earth's atmosphere, altering our climate with the potential for major social, economic and political upheaval. The world is already feeling the

[1] Nobel lecture, Oslo, December 10, 2007.

> powerful effects of global warming, and far more dire consequences are predicted if we let the growing deluge of greenhouse gas emissions continue, and wreak havoc with God's creation. A group of senior retired military officers recently warned about the potential upheaval caused by conflicts over water, arable land and other natural resources under strain from a warming planet. The problem isn't a Hollywood invention nor is doing something about it a vanity of Cassandra-like hysterics. It is a serious and urgent economic, environmental and national security challenge."[2]

Former UK Prime Minister Tony Blair and Netherlands Prime Minister Jan Peter Balkenende:

> "The science of climate change has never been clearer. Without further action, scientists now estimate we may be heading for temperature rises of at least 3–4 °C above pre-industrial levels. We have a window of only 10–15 years to take the steps we need to avoid crossing catastrophic tipping points. These would have serious consequences for our economic growth prospects, the safety of our people and the supply of resources, most notably energy. So we must act quickly."[3]

United Nations Secretary-General Ban Ki-moon:

> "We are gathered together in Bali to address the defining challenge of our age. We gather because the time for equivocation is over. The science is clear. Climate change is happening. The impact is real. The time to act is now."[4]

US Senator James Inhofe:

> "Anyone who pays even cursory attention to the issue understands that scientists vigorously disagree over whether human activities are responsible for global warming, or whether those activities will precipitate natural disasters. ... With all of the hysteria, all of the fear, all of the phony science, could it be that man-made global warming is the greatest hoax ever perpetrated on the American people? It sure sounds like it."[5]

[2] Speech on Energy Policy, April 23, 2007.

[3] Letter to Matti Vanhanen (Prime Minister of Finland and President of the EU Council), October 20, 2006.

[4] Opening speech to Bali conference on climate change, December 12, 2007.

[5] "The Science of Climate Change," floor statement by Senator James M. Inhofe, July 28, 2003.

"In addition, something that the media almost never addresses are the holes in the theory that CO_2 has been the driving force in global warming. Alarmists fail to adequately explain why temperatures began warming at the end of the Little Ice Age in about 1850, long before man-made CO_2 emissions could have impacted the climate. Then about 1940, just as man-made CO_2 emissions rose sharply, the temperatures began a decline that lasted until the 1970s, prompting the media and many scientists to fear a coming ice age. Let me repeat, temperatures got colder after CO_2 emissions exploded. If CO_2 is the driving force of global climate change, why do so many in the media ignore the many skeptical scientists who cite these rather obvious inconvenient truths?"[6]

"While the dissenting scientists (…) hold a diverse range of views, they generally rally around several key points. 1) The Earth is currently well within natural climate variability. 2) Almost all climate fear is generated by unproven computer model predictions. 3) An abundance of peer-reviewed studies continue to debunk rising CO_2 fears, and 4) "Consensus" has been manufactured for political, not scientific purposes."[7]

Professor Richard Lindzen of the Massachusetts Institute of Technology:

"Ambiguous scientific statements about climate are hyped by those with a vested interest in alarm, thus raising the political stakes for policy makers who provide funds for more science research to feed more alarm to increase the political stakes. After all, who puts money into science – whether for AIDS, or space, or climate – where there is nothing really alarming? Indeed, the success of climate alarmism can be counted in the increased federal spending on climate research from a few hundred million dollars pre-1990 to $1.7 billion today. It can also be seen in heightened spending on solar, wind, hydrogen, ethanol and clean coal technologies, as well as on other energy-investment decisions.

"But there is a more sinister side to this feeding frenzy. Scientists who dissent from the alarmism have seen their grant funds disappear, their work derided, and themselves libeled as industry stooges, scientific

[6] "Hot & Cold Media Spin Cycle: A Challenge to Journalists Who Cover Global Warming." Senate Floor Speech, Sen. Inhofe, October 25, 2006.

[7] "Global Warming 'Consensus' in Freefall," Senate Floor speech, January 8, 2009.

> hacks or worse. Consequently, lies about climate change gain credence even when they fly in the face of the science that supposedly is their basis."[8]

Professor Roy Spencer of the University of Alabama:

> "For those scientists who value their scientific reputations, I would advise that they distance themselves from politically-motivated claims of a 'scientific consensus' on the causes of global warming – before it is too late. Don't let five Norwegians on the Nobel Prize committee be the arbiters of what is good science."[9]

And Vaclav Klaus, President of the Czech Republic:

> "As someone who lived under communism for most of his life, I feel obliged to say that I see the biggest threat to freedom, democracy, the market economy and prosperity now in ambitious environmentalism, not in communism. This ideology wants to replace the free and spontaneous evolution of mankind by a sort of central (now global) planning."[10]

One of the most striking aspects of this debate is the intensity of disagreements expressed over what we might expect to be simple matters of scientific fact, such as whether the Earth is warming and whether human emissions are responsible. Such heated public confrontation over the state of scientific knowledge and uncertainty – not just between political figures and policy advocates, but also between scientists – understandably leaves many concerned citizens confused.

Our goal in this book is to clarify the climate-change debate. We seek to help the concerned, non-expert citizen to understand what is known about climate change, and how confidently it is known, in order to develop an informed opinion of what should be done about the issue. We will summarize the state of knowledge and uncertainty on key points of climate science, and examine how some of the prominent claims being advanced in the policy debate – including some in the quotes above – stand up in light of present knowledge. Can we confidently state that some of these claims are simply right and others simply wrong, or are these points of genuine uncertainty or legitimate differences of interpretation?

[8] "Climate of Fear: Global warming alarmists intimidate dissenting scientists into silence," op-ed, *Wall Street Journal*, April 12, 2006.

[9] "Hey, Nobel prize winners, answer me this," Heartland Institute, March 15, 2008, at www.globalwarmingheartland.org/Article.cfm?artId=23004.

[10] "Freedom, not climate, at risk," op-ed, *Financial Times*, June 13, 2007.

We also summarize present understanding of the likely impacts of climate change and the technologies, policies, and other options available to deal with the issue. These are not purely scientific questions, although they can be informed by scientific knowledge. In addition, we examine how scientific argument and political controversy interact. This will help to illuminate why scientific arguments play such a prominent role in policy debate over climate change, and in particular how such extreme disagreements can arise on points that would appear to be matters of scientific knowledge. What do policy advocates hope to achieve by arguing in public over scientific points, when most of them – like most citizens – lack the knowledge and training to evaluate these claims? Why do senior political figures appear to disagree on basic scientific questions when they have ready access to scientific experts and advisors to clarify these for them? And finally, what are the effects of such blending of scientific and political arguments on the policy-making process?

While there is plenty of room for honest, well-informed disagreement over what to do about global climate change, it is our view that the issue is made vastly more confused and contentious than it need be by misrepresentations of the state of scientific knowledge in policy debate, and by misunderstandings and misrepresentations of the extent of uncertainty on key scientific points about climate change and the significance of these uncertainties for action.

Before we can engage these questions, the next two sections of this chapter provide some necessary background. Section 1.2 provides a brief scientific background and primer on the Earth's climate, the greenhouse effect, climate models, and how human activities have increased greenhouse gases in the atmosphere. Section 1.3 provides a brief history of existing policy and institutions concerned with climate change, to provide the policy context for the present debate.

1.2 Climate and climate change: a scientific primer

1.2.1 What is climate?

The climate of a place, a region, or the Earth as a whole, is the average over time of the meteorological conditions that occur there – the average weather. For example, in the month of November between 1971 and 2000, the average daily high temperature in Washington, DC was 14°C, the average daily low was 1°C, and 0.3 cm of precipitation fell. These average values, along with averages of other meteorological quantities such as humidity, wind speed, cloudiness, and snow and ice coverage, define the November climate of Washington over this period.

While climate consists of average meteorological conditions, weather consists of meteorological conditions at a particular time. For example, on November 29, 1999, in Washington, DC, the high temperature was 5°C, the low was –3°C, and no precipitation fell. On this particular November day, the weather in Washington was somewhat colder and drier than Washington's average November climate.

Weather matters for short-term, day-to-day decisions. Should you take an umbrella when you go out tomorrow? Will frost kill plants left outdoors tonight? Is this a good weekend to go skiing in the mountains? Should you plan your party this weekend indoors or outdoors? In each of these cases, you care about conditions on a particular day, not long-term average conditions – the weather, not the climate.

Climate matters for longer-term decisions. If you run an electric utility, you care about the climate because if average summer temperatures increase, people will run their air conditioners more and you may need to build more generating plants to meet the increased electrical demand. If you are a city official, you care about the climate because urban water supplies usually come from reservoirs fed by rain or snow. Changes in average temperature or the timing or amount of precipitation could change both the supply and the demand for water. If the climate changes, the city may need to expand capacity to store or transport water, find new supplies, or develop policies to limit water use in times of scarcity. In Section 1.2.6 below, we will return to the difference between weather and climate, in discussing differences in their predictability.

1.2.2 Electromagnetic radiation

To understand how climate can change, we must first consider why the climate is the way it is, in particular places and for the Earth as a whole. Scientists have been studying these questions since the early nineteenth century, starting with the largest question of all: why is the Earth the temperature that it is?

The source of energy for the Earth's climate is sunlight, which is a form of *electromagnetic radiation*. Electromagnetic radiation includes all light that we can see, as well as other radiation, other light, that we cannot. Electromagnetic radiation consists of a stream of *photons*, tiny discrete packages of energy. Every photon has a size, or *wavelength*, that determines how it interacts with material in the world. Most photons emitted by the Sun have wavelength between about 0.3 and 0.8 microns.[11] This is also the range of wavelengths that are visible to

[11] A micron, or micrometer, is one one-millionth of a meter or one one-thousandth of a millimeter. A millimeter is about the width of one letter in this footnote.

human eyes. Our eyes, and those of other animals, have evolved to be sensitive to these wavelengths because of the survival advantages of being able to see the radiation that is most strongly present in the environment. Within the visible range, humans perceive different wavelengths as color. We see wavelengths near 0.3 microns as violet. As the wavelength increases, the perceived color changes to indigo, then blue, green, yellow, orange, and finally red at wavelengths around 0.8 microns. Photons with longer wavelengths, beyond red, are called *infrared* and are not visible to humans.

Most electromagnetic radiation in the universe comes from matter, through a process called *blackbody radiation*. Blackbody radiation is ubiquitous. Virtually everything in the universe, and all objects in everyday life, are constantly emitting photons. In fact, you are emitting photons right now, as is everything around you: the walls, your desk, your dog, this book. Everything is glowing.

But if everything around you is emitting radiation, why don't you see it glowing? The answer can be seen in Figure 1.1, which shows the distribution of wavelengths of the photons emitted by objects at three different temperatures. For an object at room temperature, about 20°C, almost all photons are emitted at wavelengths longer than 4 microns. These infrared photons are detectable by infrared cameras and night-vision goggles, but cannot be seen by human eyes.[12]

As an object's temperature increases, the amount of energy it emits as blackbody radiation increases. The relation between temperature and total radiated energy, known as the Stefan-Boltzmann Law,[13] states that energy emitted is proportional to the fourth power of temperature. So if the temperature of an object doubles, the rate of energy emitted increases by a factor of 2^4 or 16. This means that an object at 5600°C, like the Sun, radiates energy more than a hundred thousand times faster than an object at 20°C.

But as Figure 1.1 shows, this higher rate of radiation does not just come from emitting more photons of the same wavelengths: as an object warms up, the mix of photons it emits also shifts toward shorter wavelengths. For an object at 2200°C (middle panel, Figure 1.1), about the temperature of a piece of iron being worked by a blacksmith, most emitted photons have wavelengths too long for human eyes to see, but a few fall in the visible range. These visible photons are

[12] This explains the term "blackbody." A blackbody is an idealized object that absorbs all photons that fall on it, and emits photons with wavelengths that are determined by its temperature. At room temperature, such an object would appear to be black to the human eye.

[13] Power radiated (energy per second) per unit area is equal to σT^4, where σ is a constant (5.67×10^{-8} W/m^2/K^4) and temperature is measured in degrees Kelvin, degrees above absolute zero, which is equal to the Celsius temperature plus 273.15.

at the red end of the visible range, so the iron has a faint red glow: it has become "red hot." Blacksmiths use this to tell when a piece of metal has become hot enough to work, and the need to see this faint red glow is one reason blacksmiths often work in dim light. The Sun is, to a good approximation, a 5600°C blackbody. The bottom panel of Figure 1.1 shows that most photons emitted by a body at this temperature lie in the range that is visible to humans.

When you see a room-temperature object like this book, you are *not* seeing blackbody photons emitted by the book, because those photons are outside the visible range in the infrared. Rather, you are seeing photons that were emitted by some much hotter blackbody, the Sun or a light bulb filament (~2700°C), which have hit the page and reflected to your eye.

1.2.3 The Earth's energy balance

Photons of any wavelength are little bundles of energy. So when an object emits a photon, the photon carries a tiny bit of energy away from the object. And when a photon falls on an object and is absorbed, the object gains the photon's tiny bit of energy. Most objects – including you and everything around you – are continuously emitting photons by blackbody radiation, and at the same time absorbing photons that were emitted by other objects.

If an object is losing more energy by emitting photons than it is receiving by absorbing photons, its energy must be decreasing. Since temperature is a measure of an object's energy, this imbalance in energy emitted and absorbed causes the object's temperature to fall. Similarly, if an object is gaining more energy by absorbing photons than it is losing by emitting them, its temperature must rise. If the rates of energy gain from absorption and loss from emission are equal, the object's temperature is constant: it is in equilibrium, or steady-state.

Nearly all the photons striking the Earth come from the Sun. The amount of solar energy striking the Earth per second is truly awesome: 154 thousand trillion watts, or an average of 342 watts per square meter averaged over the whole Earth's surface. Of this, about 30 percent is reflected back to space by clouds, ice, snow, and other light-colored surfaces, so about 240 watts per square meter is absorbed by the Earth's surface and atmosphere.

In the early nineteenth century, mathematician Joseph Fourier asked a seemingly simple question: since the Earth is always absorbing energy from the Sun, why does it not heat up until it is as hot as the Sun? Blackbody radiation provides the answer to Fourier's question: the Earth and atmosphere (a rather large blackbody) radiate energy out to space, also at a rate of about 240 watts per square meter, precisely offsetting the energy absorbed from sunlight. We can use this equilibrium to estimate what the surface temperature of the Earth

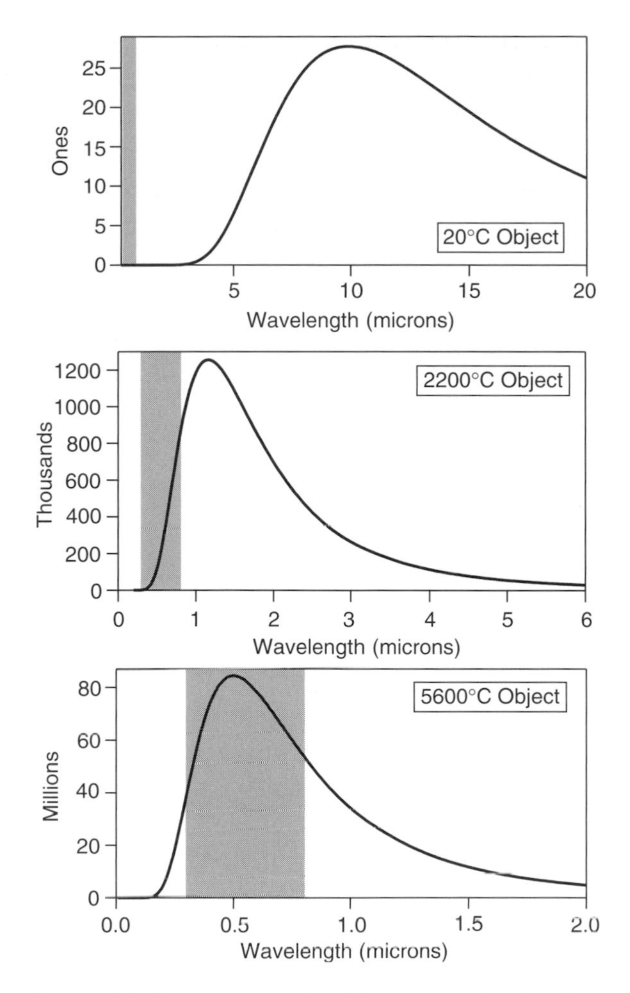

Figure 1.1 Photons emitted at different wavelengths, from objects at three temperatures. The vertical axes are (top to bottom) in units of one, one thousand, and one million watts per square meter of radiating surface, per micron of wavelength range. The gray bar in each panel shows the wavelength range visible to human eyes.

should be. Figure 1.2 plots the Stefan-Boltzmann relationship introduced above, to show how total energy emitted by a blackbody varies with temperature. Figure 1.2 shows that to radiate 240 watts per square meter, and so be in equilibrium with incoming solar radiation, the Earth must have a surface temperature of about –18°C.

This is awfully cold. Fortunately, it is also wrong. The Earth's surface is much warmer than this, a pleasant 15°C on average . The error in the calculation came from assuming the infrared radiation emitted by the surface escapes directly to space. This assumption would be correct if the Earth had no atmosphere, but

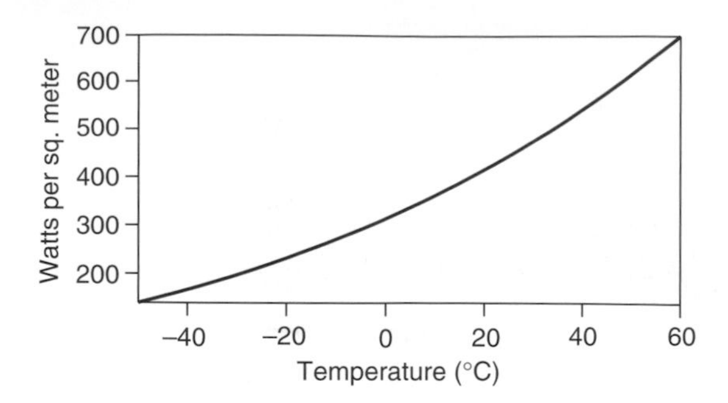

Figure 1.2 Total power radiated by a blackbody as a function of temperature (watts per square meter).

the Earth's atmosphere absorbs infrared radiation quite strongly, warming the surface by what is known as the "greenhouse effect."

1.2.4 The greenhouse effect

Recall that because the Sun is approximately a 5600°C blackbody, the photons in sunlight mostly have wavelengths around 0.5 microns. The Earth's atmosphere is essentially transparent to these visible photons, so those that are not reflected pass through it and are absorbed by the surface. Because the Earth's surface and atmosphere are much colder than the Sun, the photons they emit are in the infrared range, with wavelengths around 10 microns. The atmosphere is not transparent to these photons, but absorbs them quite efficiently.

The atmosphere is also a blackbody, albeit a gaseous one, and is also subject to energy balance: energy in equals energy out. For the atmosphere to be in steady-state, it must emit energy equal to what it absorbs from infrared photons emitted by the surface. Because blackbodies radiate in all directions, half the photons emitted by the atmosphere, on average, will go upward toward space, the other half downward toward the surface. Photons emitted down toward the surface are reabsorbed, either by the surface or the atmosphere.

We can construct a highly simplified model of the climate, in which the atmosphere consists of a single absorbing and emitting layer, and loss of energy to space comes entirely from the atmosphere, not the surface. To balance energy input from the Sun, the atmosphere would have to radiate 240 W/m^2 up into space. But since the atmosphere radiates equally upward and downward, it must also radiate 240 W/m^2 down toward the surface. This is shown schematically in Figure 1.3.

The flow of energy from the atmosphere to the surface is what we mean by the "greenhouse effect". The surface is being heated not just by visible radiation from

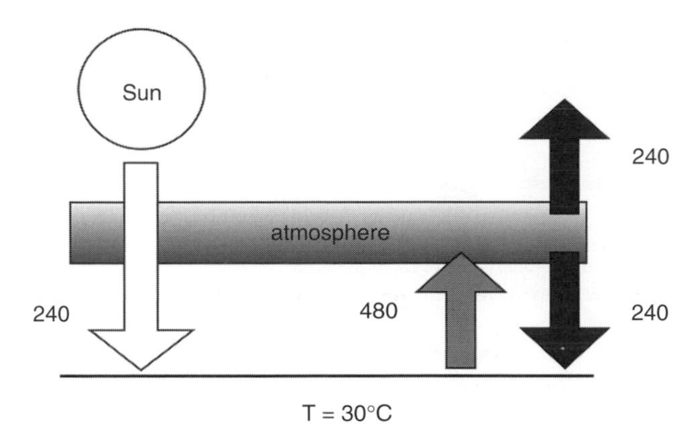

Figure 1.3 Simplified energy flow in a planet with an Earth-like atmosphere. The white arrow on the left represents incoming solar energy, the middle arrow photons emitted by the surface, and the rightmost, black arrows photons emitted by the atmosphere, in watts per square meter.

the Sun, but also by infrared radiation emitted by the atmosphere. And since the surface is receiving total energy input of 480 watts per square meter – 240 watts from the Sun and 240 from the atmosphere – the surface must also emit 480 watts per square meter to be in equilibrium. Figure 1.2 shows that to radiate 480 watts per square meter, the surface must have a temperature of about 30°C, about 48°C warmer than the surface temperature with no greenhouse effect.

This simple model shows how an atmosphere that absorbs infrared radiation can warm the surface. Remember, however, that the Earth's average surface temperature is about 15°C. This simple model gives too much warming, for two reasons. First, the atmosphere is not perfectly effective at trapping infrared radiation. It absorbs some infrared wavelengths strongly, some weakly, and some not at all, so some radiation emitted by the surface does escape directly to space. Second, physical processes that occur in the real atmosphere, such as convection – vertical stirring of the atmosphere by storms and other weather processes – serve to cool the surface. Still, the basic physics of this model is correct: the aggregate effect of infrared absorption by the atmosphere is to warm the Earth's surface above what it would be without an atmosphere.

Figure 1.4 shows a second simple model, of an Earth with a thicker, more absorbing atmosphere that we represent by two separate layers. Photons emitted by the surface are absorbed by the lower atmospheric layer. Photons emitted upward by the lower layer are absorbed by the upper layer. Photons emitted upward by the upper layer escape to space, while those emitted downward by the upper layer are absorbed by the lower layer.

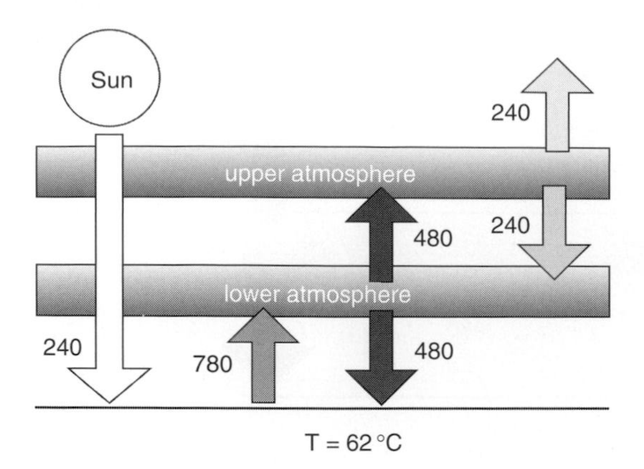

Figure 1.4 Simplified energy flow in a planet with a thicker atmosphere and stronger infrared absorption. The white arrow on the left represents incoming solar energy. The other arrows, from right to left, represent photons emitted by the lower atmosphere, upper atmosphere, and surface. Values are in watts per square meter.

In this case, only one-third of the photons emitted by the surface eventually escape to space. The other two-thirds are trapped in the atmosphere and eventually reabsorbed by the surface. Since total emission to space must be 240 watts per square meter to balance incoming sunlight, the surface must therefore be emitting at three times this rate, or 720 watts per square meter. Figure 1.2 shows that to emit 720 watts per square meter, the surface must have a temperature of 62°C. A thicker atmosphere that absorbs more infrared radiation makes the surface hotter.

Ever since this warming of the surface by the atmosphere was first described in the nineteenth century, it has been widely called the “greenhouse effect.” More recently, it has been compared to wrapping a blanket around the Earth. Neither of these analogies is really accurate, however, since both blankets and greenhouses mainly work by slowing the physical escape of warm air rather than by disrupting the passage of radiation to space.

Merely having an atmosphere does not produce a greenhouse effect: the atmosphere must absorb infrared radiation. On Earth, 96 to 99 percent of the atmosphere consists of molecular nitrogen (N_2) and oxygen (O_2) and the inert gas argon (Ar). These simple molecules do not absorb or emit infrared photons, so they generate no greenhouse effect and do not warm the surface. Rather, the greenhouse effect is caused by several other constituents whose more complex molecular shapes cause them to absorb infrared radiation: primarily water vapor and carbon dioxide, along with a few others.

Water vapor (H_2O) is the most important greenhouse gas in the atmosphere, responsible for about two-thirds of the greenhouse effect. Water vapor’s abundance is highly variable. Near the surface in the humid tropics it may reach 4 percent, while in cold Polar regions it can be just a fraction of a percent. In the stratosphere, its typical abundance is only 0.0005 percent (5 parts per million or ppm). Oxygen, nitrogen, water vapor, and argon make up more than 99.95 percent of the atmosphere. But the many species that make up the remaining 0.05 percent, or 500 ppm, play a surprisingly important role in atmospheric pollution and climate, including a large contribution to the greenhouse effect. Carbon dioxide (CO_2) is the largest share of this remainder, about 385 ppm today, and the largest contributor after water vapor to the greenhouse effect. Methane (CH_4), the next largest contributor, absorbs infrared radiation about 20 times more strongly than CO_2 on a per molecule basis, but is present in the atmosphere at only about 1.75 ppm. Smaller greenhouse contributions come from nitrous oxide, N_2O, present at about 0.3 ppm; the chlorofluorocarbons (CFCs) and related synthetic chemicals, present at about 0.003 ppm; and ozone (O_3), found in varying abundances ranging from as low as 0.01 ppm near the ground to 10 ppm in the stratosphere.

Naturally occurring greenhouse gases, most importantly water vapor and CO_2, warm the Earth's surface to its present comfortable state. And it is clear that increasing the concentrations of these gases could make the Earth warmer still. This possibility was proposed by scientists more than a century ago, first by the Swedish chemist Svante Arrhenius in 1896, and again with more supporting evidence by the British engineer Guy Callendar in 1938. The models represented in Figures 1.3 and 1.4 show how it works in highly simplified terms. Figure 1.3 shows an atmosphere whose infrared absorption is similar to the Earth's, while Figure 1.4 shows a planet with more greenhouse gases in the atmosphere. As the quantity of greenhouse gases in the atmosphere increases, less of the energy emitted by the surface escapes into space so the surface gets warmer.

An extreme example of this is the planet Venus. Venus's atmosphere contains about 250 thousand times more CO_2 than the Earth's atmosphere, so it is extremely effective at trapping infrared photons. If we used the scheme of Figures 1.3 and 1.4 to represent Venus's atmosphere, we would need about 100 layers, so a photon would have to be absorbed and re-emitted upward 100 times before escaping from the atmosphere. Consequently, only 1 percent of the photons emitted by the surface make it through the atmosphere; the other 99 percent are trapped and eventually reabsorbed by the surface. Venus absorbs incoming solar energy at a rate of about 180 watts per square meter,[14]

[14] This is a lower value than for Earth. Despite being closer to the Sun, Venus is covered by a thick shroud of clouds, which reflect a far higher percentage (about 70 percent) of the Sun's photons than the Earth does. This increased reflectivity lowers absorbed energy more than the decreased distance to the Sun increases it. As a result, absorbed energy from the Sun is lower on Venus than the Earth.

so the planet must emit back to space at the same rate to be in equilibrium. But because only 1 percent of the photons emitted by the surface escape to space, the surface must emit at one hundred times that rate, about 18,000 watts per square meter. This in turn requires a surface temperature of 480°C – hot enough to melt lead, hotter even than Mercury, which is only half as far from the Sun as Venus but has no atmosphere to provide a greenhouse effect.

1.2.5 Feedbacks and climate sensitivity

Even the simplest models, like those above, show that the Earth's surface must warm if the amount of CO_2 or other infrared-absorbing gases in the atmosphere is increased. But how much will it warm? How sensitive are the global surface temperature and the global climate to increased greenhouse gases? Answering this quantitative question requires more complex calculations than those above, which incorporate many additional processes operating in the real atmosphere.

Scientists have been doing these calculations for nearly a hundred years, first by hand and now for several decades with computers. To provide a simple basis for comparison, many of these calculations express their results in terms of how much the Earth would warm from doubling CO_2. For many centuries before the industrial revolution, CO_2 was present in the atmosphere at about 280 ppm: doubling its concentration would bring it to 560 ppm. Even if such a doubling occurred instantly, the Earth would not warm up immediately. It would take centuries or longer to reach the new equilibrium temperature because the oceans, with their enormous heat capacity, would slow the warming. The warming that eventually results from this doubling, once a new equilibrium is achieved, is called the climate sensitivity. We can speak of the climate sensitivity of a model or calculation, which we observe by running the model, or the climate sensitivity of the Earth – which we can estimate through model calculations or records of past climate fluctuation, but cannot observe directly.

If CO_2 is doubled and nothing else in the climate system changes, it is relatively easy to calculate that the resultant warming would be about 1°C. But in reality, it would not be possible to change only the concentration of CO_2. As the climate warms in response to the increased CO_2, many other things change. Most importantly, as the atmosphere warms it holds more water vapor. Since water vapor is also a greenhouse gas, this causes additional warming. Such knock-on effects of increased CO_2 – additional changes caused by the initial change – are called *feedbacks*, and are responsible for much of the warming caused by increasing greenhouse gases. Water vapor is the most powerful feedback, capable of doubling the warming caused by CO_2 alone, but many other feedbacks are also important in the climate system. Another example involves melting ice as the climate warms, mostly in glaciers and the oceans near the Poles. Ice is highly reflective, and the land or water exposed when ice melts are darker. Consequently, a reduction in ice increases the fraction of incoming solar energy absorbed by the Earth's surface, also causing further warming.

The water-vapor feedback and the ice feedback are both examples of *positive feedbacks* – feedbacks that amplify an initial warming. There are also *negative feedbacks*, by which the initial warming causes changes that produce cooling. For example, the temperatures of the surface and the upper atmosphere are linked by vertical mixing from thunderstorms: as the surface warms, so does the upper atmosphere. Since a warmer atmosphere radiates more energy into space, this effect will offset some of the warming caused by increased greenhouse gases.

Including all known feedbacks, current estimates are that doubling CO_2 from 280 to 560 ppm would lead to equilibrium warming of 2.0 to 4.5°C, with a best estimate of 3°C. In other words, feedbacks roughly triple the direct warming effect of greenhouse gases.

1.2.6 Climate models and weather models

How are these quantitative estimates of climate sensitivity generated? Other scientific fields can use controlled experiments to study the behavior of things they are interested in – e.g., atoms, fluids, or bacteria – but climate scientists cannot conduct controlled experiments on the Earth to observe how it responds to changes in atmospheric composition. Instead they use *Global Climate Models*, or GCMs – mathematical representations of the Earth that run on computers. These models represent the known physical laws that govern the behavior of the climate system – such as conservation of energy, momentum, and mass – as well as evaporation of surface water, condensation of water in the atmosphere to form clouds, and many other known physical processes and feedbacks. To provide accurate representations of the climate, GCMs must also represent the behavior of other parts of the Earth that interact with the climate, including the oceans, the land surface, the cryosphere (surface ice), and the biosphere (the world's ecosystems).

The biggest challenge to producing an accurate GCM comes from the vast range of spatial scales at which atmospheric processes operate – from pressure systems of thousands of kilometers, to clouds of a few kilometers, to turbulent eddies of a few meters, to molecular activity of millionths or even billionths of a meter. Models must divide the atmosphere into finite sized grid-cells, which are the smallest units for which they explicitly define and calculate atmospheric properties. Present computing speeds limit the smallest atmospheric grid-cells to about 100 kilometers horizontally, sliced into vertical layers about one kilometer thick. Processes operating at smaller scales than this, such as clouds, cannot be represented explicitly in the models but must instead be *parameterized.* Parameterization means representing the effects of these smaller-scale processes as functions of variables the model does explicitly resolve, such as temperature and water vapor. So while GCMs cannot represent individual clouds, which are much smaller than a single grid-cell, they can estimate the average cloudiness of a cell as a function of the cell's relative humidity and winds. Parameterizations are highly diverse. Some have well-founded physical bases, while others are *ad hoc* constructions that let the model produce a realistic present-day climate. Consequently, parameterizations are one of the largest sources of uncertainty in GCMs.

GCMs can be tested by examining how well they reproduce the Earth's actual climate. For example, a GCM might be started with conditions as they stood a few hundred years ago, then run through the present to see how well it reproduces the observed climate record. Other tests can include checking how well a model reproduces the observed climate response to some known disruption like a large volcanic eruption or an El Niño cycle. Present GCMs do pretty well on these tests. They reproduce the observed climate of the past couple of centuries quite well for the global average and for large continental regions, but agree less well with the historical record (and with each other) for smaller-scale regions. Variables that rely primarily on large-scale processes, like temperature, are better simulated than variables like rainfall that rely more on small-scale processes.

GCMs can also be used to study the climate in ways in which it is impossible to study the real Earth. They can examine "what-if" scenarios, e.g., what if the output of the Sun changed, or what if there were no human emissions of CO_2. They can also, as Chapter 3 will discuss, project how climate will change in response to specified future trends in emissions of CO_2 and other greenhouse gases.

GCMs are similar to the atmospheric models used to forecast the weather, but differ from them in important ways. The weather tomorrow depends on the weather today, so using a model to forecast the weather requires detailed information on the state of the atmosphere today: the temperature, pressure, humidity, and many other characteristics, measured as precisely and in as many places as possible, both on the surface and at higher altitudes. A weather forecasting model takes this snapshot of the atmosphere and uses known physics to move it forward in time to predict how the state of the atmosphere will evolve over time. But the initial snapshot of the atmosphere is never perfectly accurate or complete, and because the physical relations determining how the atmosphere evolves are non-linear, small errors in the starting description grow exponentially with time. Within a week or two, the errors dominate and the model's forecast is essentially useless.

Given this fundamental limit to models' ability to predict the weather, how can model projections of the climate several decades in the future possibly be accurate? The answer lies in the differences between weather and climate. As discussed in Section 1.2.1 above, weather is the state of the atmosphere at a given moment, while climate is the statistics of the weather over some time period. Limited ability to predict the weather on a particular day does not imply limited ability to predict the distribution of conditions over time – e.g., how warm is an average day, and how hot or cold and how frequent are the most extreme days – that make up the climate. These are different quantities that pose different prediction problems, and the prediction of averages and distributions is easier – just as it is easier to predict the distribution of outcomes on repeated tosses of a coin than it is to predict the outcome of a single toss. Short-term unpredictability of specific conditions does not imply long-term unpredictability of average conditions and distributions.

Consider another example. Suppose it is January in Ann Arbor, Michigan. Can we predict whether the average temperature next June will be warmer or colder than this month's average? Of course we can: we know with certainty that June will be warmer than January, even if we cannot predict the weather for June 15. June is warmer because the northern hemisphere, including Ann Arbor, is tilted toward the Sun in June, so it receives more sunlight than in January. With more solar input, it must be warmer for outgoing infrared radiation to balance the incoming energy.[15] Similarly, increases in greenhouse gases increase the heating of the surface by the atmosphere. This in turn requires the surface to warm for energy balance to be achieved. Thus, we need not be able to predict the weather on any day in the 2080s to know that if atmospheric greenhouse gases have significantly increased at that time, that decade will be warmer than this decade.

This is not to say predicting climate is easy. While climate predictions do not suffer from the uncertainty about initial conditions that limits weather forecasts, they have other problems that weather forecasts do not. For example, over the century or more of climate forecasts, many components of the climate system may change in response to warming, such as the distribution of ice sheets, the circulation of the oceans, and the distribution of ecosystem types over the

[15] The simple models above considered the average balance of incoming and outgoing energy over the whole globe, which must be equal for equilibrium. A single location is different: its balance of incoming and outgoing radiation need not be equal, because it can also lose energy by horizontal transport to nearby places. This qualification does not matter for the large-scale seasonal cycle discussed here, however.

land surface. Weather forecasts can ignore such changes, because these parts of the climate system do not change over the one or two weeks that a weather forecast covers. But a climate forecast must consider how these things may change, and interact with other elements of climate, over decades or longer. This is one of many uncertainties in climate forecasting, which we discuss in more detail in Chapter 3. But these uncertainties concern how much warming there will be, how it will be distributed around the Earth, and other matters - not whether it will warm.

1.2.7 Increased climate forcing from human activities

Over the past two centuries, human activities have sharply increased the atmospheric abundance of several greenhouse gases. The most important increase has been CO_2, which is emitted from burning fossil-fuel energy sources - coal, oil, and natural gas - and from land clearing and deforestation. These CO_2 emissions from human activities are superimposed on a natural global carbon cycle, by which CO_2 is constantly exchanged between the atmosphere and ecosystems as organisms take up CO_2 through photosynthesis and release it through respiration, and between the atmosphere and oceans via both physical and biological processes. Figure 1.5 shows how the atmospheric abundance of CO_2 has varied over the past 10,000 years. It remained between 260 and 280 ppm for most of that period, then began a rapid increase in the nineteenth century, closely tracking the rise in fossil-fuel combustion in the industrial revolution. Atmospheric CO_2 now stands just above 385 ppm, and is increasing about 2 ppm per year.

Methane (CH_4), also known as natural gas, is the second most important greenhouse gas emitted by human activities. Methane is emitted from rice paddies, landfills, livestock, biomass burning, and the extraction and processing of fossil fuels, as well as several natural sources. The time series of its atmospheric abundance looks similar to that of CO_2: before 1800, it was constant at about 0.8 ppm; starting in 1800 it began a rapid rise; and today it has more than doubled to about 1.75 ppm. Other greenhouse gases increasing due to human activities include nitrous oxide (N_2O), which is emitted from nitrogen-based fertilizer and industrial processes plus a few natural sources and has increased from about 250 to 315 parts per billion (ppb); and the halocarbons, a group of synthetic chemicals of which the most important are the chlorofluorocarbons (CFCs), used as refrigerants, solvents, and in various other industrial applications. These chemicals have no natural sources, but are now present in the atmosphere at about 3 ppb. Together with several additional smaller human influences, the elevated concentration of these greenhouse gases is now responsible for increasing the

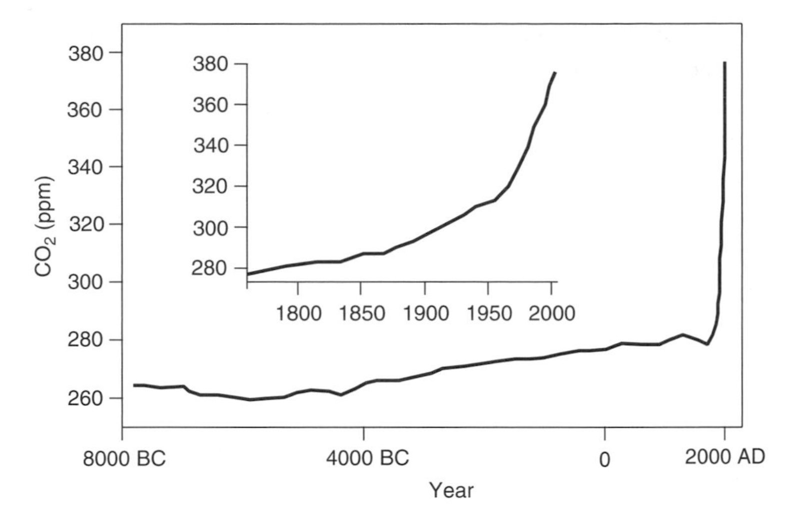

Figure 1.5 Global average concentration of CO_2 in the atmosphere over the past 10,000 years, in parts per million (ppm). The inset figure shows the last 250 years. *Source*: Figure SPM.1, IPCC [2007a].

energy input to the Earth's surface by about 1.6 watts per square meter, relative to the solar input of 240 watts.

Knowledge of these human influences on the climate has grown over the past several decades. By the 1950s and early 1960s it was becoming clear that fossil-fuel use was releasing CO_2 fast enough to significantly increase its atmospheric abundance, and the first careful measurements of this increase were published in the 1960s. While most scientists expected the climate to warm in response to this, there was also some evidence suggesting the possibility of a cooling trend, including the record of past climate oscillations between ice ages and warm interglacial periods. The current warm period has lasted about 10,000 years, roughly the length of previous warm interglacial periods. This might suggest the Earth is due for a gradual, long-term cooling leading to another ice age. Moreover, global temperatures showed a slight cooling trend between about 1945 and 1975, at the same time as it was becoming clear that smoke and dust pollution could shade the Earth's surface from incoming sunlight and so magnify any natural cooling trend. By the early 1980s, however, global temperatures had resumed warming and many new pieces of evidence showed that warming from greenhouse gases was the predominant concern.

With elevated concentrations of greenhouse gases in the atmosphere, we expect the Earth to warm and the climate to change. As Chapter 3 will discuss, the best present projections are that if emissions of CO_2 and other greenhouse gases keep growing more or less as they have been, the Earth will warm about

3 degrees Celsius by the end of this century. This might not sound like much, since even in one place the difference between a hot summer day and a cold winter one can be as large as 50°C, and changes half that large can occur from day to night or one day to the next. Therefore, you might reasonably guess that a global temperature increase of a few degrees is not likely to matter much – but this would be a serious error. While the temperature of any single place on the Earth can vary greatly, the average temperature of the whole Earth is quite constant, through the year and from year to year. In the Earth's past, changes of only a few degrees in global-average temperature have been associated with extreme changes in climate. For example, at the peak of the last ice age 20,000 years ago – when glaciers thousands of feet thick covered most of North America – the average temperature of the Earth was 5–8°C cooler than it is today. Thus, the prospect of a few degrees Celsius rise in global temperature over just 100 years – and more beyond – must be viewed with utmost seriousness. In Chapter 3 we will summarize what has been learned since climate change emerged as a serious scientific question nearly 50 years ago, about the evidence that the climate is changing, the attribution of those changes to human activities, the likely future changes, and their impacts.

1.3 Background on climate-change policy

Like many serious environmental issues, global climate change came to the attention of policy-makers after decades of related scientific research. Climate change attracted virtually no public or political attention in the 1960s, and only a little during the energy-policy debates of the 1970s. By the early 1980s, as it became increasingly clear that warming from greenhouse gases was a serious concern, scientists and scientific organizations began trying to persuade governments to pay attention to the climate problem. They had little success until 1988, when several events brought climate change suddenly onto the political agenda.

That summer, North America suffered an extreme heat wave and the worst drought since the dust-bowl years of the 1930s. By July, 45 percent of the United States was in a drought and a few prominent scientists stated publicly that global climate change was probably the cause. Moreover, this extreme summer followed a period of intense worldwide publicity about the Antarctic ozone hole and the negotiation of the Montreal Protocol, the international treaty to control the ozone-depleting chemicals. Under these conditions, politicians and the public were primed to consider the possibility that human activities could be disrupting the global climate. In late 1988, instead of naming a "Person of the Year," *Time* magazine designated "Endangered Earth" the "Planet of the Year," while

Climate change and ozone depletion

People often confuse global climate change with depletion of the stratospheric ozone layer, but these are distinct problems. Ozone is a molecule made up of three oxygen atoms, which occurs naturally in the stratosphere, about 15 to 40 kilometers above the Earth's surface. Ozone in the stratosphere protects life on Earth by absorbing most of the highest-energy ultraviolet (UV) radiation in sunlight. Although most sunlight lies in the visible range, a few percent of it has wavelengths shorter than 0.3 microns, in the ultraviolet. To make things more confusing, ozone near the ground is a health hazard and a component of smog, which human activities are increasing. To keep straight the difference between "good ozone" (up there) and "bad ozone" (down here), you need only remember that you want ozone between you and the Sun, but do not want to breathe it.

Beginning in the 1970s, scientists realized that a group of man-made chemicals, mainly the chlorofluorocarbons or CFCs, could destroy stratospheric ozone. This would cause more intense UV radiation to reach the surface, bringing increased skin cancer, cataracts, and other harms to human health and ecosystems. Concern rose further in the 1980s, when extreme ozone losses were observed over Antarctica each spring (October and November) – soon called the "ozone hole" – and CFCs were identified as the cause.

After ten years of unsuccessful attempts to solve the problem, nations adopted strict regulations in the late 1980s and 1990s that have nearly eliminated most ozone-depleting chemicals in industrialized countries. Developing countries are now phasing out the chemicals. As a result, the amount of CFCs in the atmosphere has already begun to decline, and stratospheric ozone is projected to recover gradually over 30 to 50 years.

Climate change and ozone depletion are linked in a few ways. One link is that CFCs are strong infrared absorbers, so they contribute to climate change as well as destroying ozone. Another link is that climate change warms the Earth's surface and lower atmosphere, but makes the stratosphere colder and wetter. Colder, wetter conditions promote ozone destruction, so will likely delay recovery of the ozone layer even if phaseouts of ozone-depleting chemicals stay on course. But despite these linkages, ozone depletion and climate change are fundamentally different problems. They have different causes: CFCs and other chemicals containing chlorine or bromine, versus CO_2 and other greenhouse gases. And they have different effects: stronger UV radiation harming health and ecosystems, versus changes in climate and weather worldwide. Despite the important differences between these issues, many aspects of how nations responded to ozone provide useful lessons for responding to global climate change. We will refer to relevant aspects of the ozone issue at several points in this book.

the United Nations General Assembly passed a resolution stating that the climate was "a concern to mankind."

Governments' first response was to establish an international body to assess scientific knowledge of climate change, the Intergovernmental Panel on Climate Change or IPCC. The IPCC involved hundreds of scientists organized into three working groups, each responsible for a different aspect of the climate issue: the atmospheric science of climate change; the potential impacts of climate change and ways to adapt to the changes; and the potential to reduce the greenhouse-gas emissions contributing to climate change. The four major assessment reports that the IPCC has completed since its formation, in 1990, 1995, 2001, and 2007, are widely regarded as the authoritative statements of scientific knowledge about climate change. We will refer to these assessments repeatedly throughout this book.

As the IPCC was beginning its work in the late 1980s, governments also began considering actions to respond to climate change. Over the two years following the hot summer of 1988, several high-profile international conferences called for reducing worldwide CO_2 emissions, typically by 10 to 20 percent as a first step. Through 1991 and 1992, national representatives worked to negotiate the first international treaty on climate change, the Framework Convention on Climate Change (FCCC). Signed in June 1992, this treaty entered into force in 1994 and has since been established law in all the nations that have ratified – now numbering more than 190, including the United States.[16]

The FCCC's stated objective is "Stabilization of greenhouse gas concentrations in the atmosphere at a level that would prevent dangerous anthropogenic interference with the climate system ... within a time-frame sufficient to allow ecosystems to adapt naturally to climate change, to ensure that food production is not threatened, and to enable economic development to proceed in a sustainable manner." The treaty also states several principles intended to guide subsequent

[16] After a treaty has been negotiated and signed by national representatives, it enters into force, or becomes legally binding, only after enough nations take the second step of ratifying it – formally expressing their commitment to be bound by it. Every treaty specifies how many nations must ratify for it to enter into force. After these are received, the treaty becomes a binding legal obligation, but only for those who have ratified it.

climate-policy decisions, of which a particularly important one is the principle of "Common but differentiated responsibility." This principle states that all nations have an obligation to address the climate issue, but not in the same way or at the same time, and in particular that "… the developed-country Parties should take the lead in combating climate change and the adverse effects thereof."

The FCCC was not intended to be the last word on climate change, but to provide a starting point for more specific and binding measures to be negotiated later. Consequently, in contrast to its ambitious principles and objectives, the treaty's concrete measures were weak and preliminary. Under the FCCC, nations committed to reporting their current and projected emissions and supporting climate research. Parties also accepted a general obligation to adopt, and report on, measures to limit emissions. What these measures had to be, or had to achieve, however, was not specified. Only for the industrialized countries (called "Annex 1 countries") did this general obligation also include the specific aim of returning emissions to 1990 levels by 2000. This aim was the closest the FCCC came to concrete action to advance its objectives, but even it was not legally binding.

Weak as this aim was, few governments made serious efforts to meet it. Many, including the USA, assembled national programs that were little more than exhortations for voluntary action and re-labeling of existing programs. The few nations that met the target largely did so by historical accident or through policies adopted for other reasons. Russia, for example, met the target because of the collapse of the Soviet economy after 1990, Germany because it absorbed the shrinking East German economy, and Britain because it privatized electrical generation and cut coal production. It was clear immediately after adoption of the FCCC that making significant emission reductions would require stronger measures. After a few years of wide-ranging debate about what form these stronger measures might take, a plan was adopted in 1995 to negotiate binding national greenhouse-gas emission limits for the Annex 1 (industrialized) countries. In December 1997, these negotiations concluded with the signing of the Kyoto Protocol.[17]

Negotiations of the Kyoto Protocol were marked by hard, last-minute bargaining over national emission limits. European and Japanese delegations sought ambitious cuts to bring emissions 5 to 15 percent below 1990 levels by 2010. The Clinton administration initially opposed near-term emission cuts, and instead proposed only research and voluntary initiatives in early years, with emission

[17] "Conventions" and "Protocols" are both treaties. A Convention is typically a broad agreement that provides a framework for more specific agreements negotiated in Protocols under the Convention. In this case, parties to the FCCC negotiated the Kyoto Protocol to advance the objectives and principles laid out in the FCCC.

targets coming into effect only after 2008. The US Senate took the unusual step of expressing its hostility to emission targets even before negotiation of the treaty was completed, passing a resolution (the "Byrd-Hagel resolution") that rejected emission limits for industrial countries unless developing countries made commitments to emission cuts at the same time.

The agreement reached in the final hours of the Kyoto Conference imposed specific emission targets for each industrialized country over a five-year "commitment period" from 2008 to 2012. Targets were defined for total emissions of a basket of CO_2 and five other greenhouse gases. Although the treaty included no emission limits for developing countries, the US delegation signed it, in defiance of the Senate resolution. The emission targets were 8 percent below 1990 levels for the European Union and a few other European nations; 7 percent for the United States; 6 percent for Japan and Canada; and zero (i.e. hold emissions to their 1990 level) for Russia and Ukraine.[18] If all nations met their targets, emissions from these nations would be 5.2 percent below 1990 levels over the commitment period.

The Protocol also incorporated several hastily drafted provisions to allow flexibility in how nations met their emission limits. These included mechanisms to exchange emission-reduction obligations between nations (allowing one nation to cut less by paying another country to make a larger cut). They also included provisions for nations to meet some of their obligation by enhancing carbon uptake through planting trees or similar measures, instead of reducing emissions from energy use or industry. The details of these provisions, however, along with many other matters of how to implement the Protocol, were left to be settled later.

Further negotiations over the three years following the Protocol's signing sought to establish more specific rules for implementing the emission commitments, particularly regarding how much credit nations could claim for enhancing carbon uptake and for financing emission reductions abroad under the flexibility mechanisms. These negotiations revealed sharp differences between two groups of industrialized countries over how much flexibility should be granted. One group, including the USA, Russia, Japan, Canada, and several others, sought more liberal credit for enhancing CO_2 uptake by forests or other sinks, and more flexibility to substitute cuts abroad for cuts at home, while most European nations wanted to allow less flexibility on both these points.

[18] A few smaller nations negotiated particularly advantageous commitments for themselves: New Zealand's target, like Russia's, was to hold emissions at their baseline level; Norway was allowed a 1 percent increase above their baseline; Australia an 8 percent increase; and Iceland a 10 percent increase.

This conflict came to a boil and negotiations between the two groups broke down at a conference in November 2000 in The Hague. Here, despite political shifts toward a harder line in Europe and the looming uncertainty of the unresolved US Presidential election, delegates nearly reached a compromise. But the proposed compromise was rejected at the last minute by the French and German environment ministers (both Green Party members), who judged that the proposed weakening of the Kyoto commitments was too high a price to pay for US participation. While the negotiation breakdown was widely blamed on this conflict, there were several other looming disagreements, both between industrialized and developing countries and among developing countries, that could also have blocked agreement if they had come to the top of the agenda.

While the Clinton administration was confused and inconsistent in its approach to the climate issue overall and the Kyoto Protocol specifically, the Bush administration's attitude to the Protocol was clear hostility. Two months after taking office in 2001, the new administration announced it would not ratify the Protocol, because there was too much scientific uncertainty about climate change and because the Protocol's emission limits would harm the US economy. Although it later retreated from claiming that its withdrawal was based on scientific uncertainty, the Bush administration continued to hold that the Protocol was unacceptable, because of high costs to the US economy and the absence of emission limits for developing countries. In February 2002, President Bush outlined his alternative approach to the issue, which included several components: a target of reducing the "greenhouse gas intensity" of the American economy – emissions per dollar of GDP (gross domestic product) – by 18 percent by 2012;[19] increased funding for climate-change science and for specific technologies to reduce emissions; tax incentives for renewable energy and high-efficiency vehicles; and several programs to encourage voluntary emission cuts by businesses.

Following the announced US withdrawal, other signatories continued to negotiate over the flexibility mechanisms and provisions for compliance, agreeing on a compromise in November 2001, similar to the one rejected in 2000. These agreements allowed more flexibility than European delegations had previously

[19] Note that this target is measured in emissions relative to the size of the US economy, not emissions themselves. The emission level that is allowed under this target grows with the economy, so if the economy grows more than 18 percent, total emissions under the target would increase. Further, the target rate of improvement is not particularly ambitious since it is roughly equal to the reduction in greenhouse-gas intensity that was realized during the 1990s.

been willing to accept, and were followed by announcements that the European Union, Japan, and somewhat later Canada, would ratify the Protocol.

Still, the legal force of the Protocol remained uncertain until late 2004. To enter into force – and so become binding on those who ratified – the Protocol required ratifications by 55 countries, including nations contributing at least 55 percent of 1990 industrialized-country emissions. This threshold meant that, without the United States, the treaty could enter into force only if all other major industrialized countries joined, including Russia. After several years of uncertainty about its intentions, Russia submitted its ratification in November 2004, allowing the Protocol to enter into force on February 16, 2005.

But the long delay awaiting the required ratifications meant that the Protocol entered into force only three years before the start of the commitment period. Nations' efforts in the meantime had been highly uneven, and many were not well prepared to achieve what would be a large deflection of emissions over very few years. As a result, national emissions trends have been highly uneven and few parties are likely to meet their commitments.

Russia and Ukraine have the easiest compliance situations, because the collapse of the Soviet economy in the early 1990s brought emissions far below their target, which was set at their 1990 baseline. In 2006, Russia's emissions were 34 percent below the target and Ukraine's 52 percent below, allowing them to sell credits to other nations under the Protocol's flexibility mechanisms.

The most serious efforts to reduce emissions thus far have been made by the European Union (EU). Under the Protocol, the EU must reduce emissions 8 percent, but is allowed to meet this target in aggregate, so larger cuts in some member states can offset smaller cuts or emissions growth in others. The EU has committed to much more aggressive targets, however – 20 percent below 1990 levels by 2020, to be tightened to a 30 percent cut if other industrial countries agree to make similar efforts. To achieve these cuts, the EU has implemented a system of tradable emissions permits covering power plants and most large industrial sources (about half of total emissions), plus regulations covering buildings, vehicles, appliances, other equipment, and biofuels. Despite these efforts, EU emissions in 2006 were only 2.7 percent below the 1990 baseline, with additional measures optimistically projected to bring them to 7 percent below baseline by 2010. Formal EU compliance with its 8 percent Kyoto cut consequently remains within reach, but only with substantial funding of foreign reductions under the Kyoto Protocol flexibility mechanisms.

Among other major industrial nations, the United States is not obliged to meet its 7 percent Kyoto cut because it did not ratify. US emissions were 16 percent above its baseline in 2007. Japan and Canada both made far less serious efforts than the EU, and achieved less. Japan's 2007 emissions were 9 percent above baseline, and the many small measures in its climate plan were projected, optimistically, to reduce only a few percent, so Japan is unlikely to meet its targeted 6 percent cut even with large funding of foreign reductions. Canada is in an even worse situation, having ratified only in December 2002 and proposed several different climate plans, all weak and none fully implemented. With Canada's 2007 emissions 26 percent above baseline, there is no chance of achieving the target, despite the legal obligation under the Protocol. The party in perhaps the easiest position is Australia. Originally opposed to the Protocol like the United States, Australia changed course and ratified after a change of government in 2007. But with one of the most lenient Kyoto targets (an 8 percent *increase* from baseline) and emissions growth of only 7 percent from 1990 to 2006, Australia has a fair chance of achieving its target despite the late start. Overall, emissions from nations with Kyoto commitments were 14 percent below their 1990 baseline in 2007, but this large decrease was predominantly due to the collapse of the Russian economy, and its emissions, in the early 1990s. Most Kyoto signatories with emissions targets will miss them, and several major economies will miss them by a lot.

With this failure looming, parties meeting in Bali in December 2007 agreed to negotiate a new climate treaty within two years, to include actions by both industrialized and developing countries. Subsequent negotiations quickly reverted to old lines of conflict, however, so by early 2009 the goal was reduced to negotiating another political commitment, with a treaty to follow still later. As the December 2009 Copenhagen meeting approached, hopes were raised by renewed US engagement under the Obama Administration, plus many national statements of increased commitments. The meeting was marked, however, by sharp disputes over sharing the burden of action and many procedural roadblocks. Failure was only avoided through intensive last-minute negotiations of a political deal by leaders of 28 nations, including large emitters and representatives of regional groups. This "Copenhagen Accord" included several significant advances on any prior agreement – on a goal of limiting climate change to 2 °C, on emission cuts to achieve this, on verification of national actions, and on financial support for developing countries – but was vague or weak on key points. Moreover, objections from just five nations blocked the accord from formal adoption, so its status even as a basis for future negotiations is uncertain. In sum, despite modest favorable signs before and at Copenhagen, and current high attention to climate change, it remains unclear whether leaders of major nations are willing to act strongly enough to address the problem, or whether the current international negotiation process is able to motivate and coordinate such action.

1.4 Plan of the book

The remainder of the book seeks to provide a clear guide to the present climate-change debate. It summarizes the present state of scientific knowledge about climate change, the policy options available to respond to it, the political debate over what to do about it, and how these three areas of knowledge and debate – science, policy, and politics – interact.

The plan of the book is as follows. Chapter 2 discusses the general characteristics of scientific debate and political debate, the differences between them, and the predictable challenges that arise when important questions lie on the boundary between these dissimilar areas of debate and decision-making. Chapter 3 summarizes present scientific knowledge and uncertainty about global climate change, focusing on the points that have become the most prominent matters of public controversy. Chapter 4 summarizes present knowledge about potential technological and policy responses to the climate issue. Finally, Chapter 5 does two things. First, it outlines the present political debate on climate change and the foundations of the present deadlock on the issue. Second, it draws on our own judgments to present a set of explicit recommendations of what should be done to respond appropriately to the grave threat posed by global climate change.

Further reading for Chapter 1

David Archer (2007). *Global Warming: Understanding the Forecast*. Malden, MA: Blackwell Publishing.

K. Emanuel (2007). *What We Know About Climate Change*. Cambridge, MA: MIT Press.

These two relatively short books describe the basic science of global warming. They are written for those without a deep scientific background.

2

Science, politics, and science in politics

The climate-change debate, like all policy debates, is fundamentally an argument over action. How shall we respond to climate change? Do the risks it poses call for action, and if so, how much effort – and money – shall we expend, and on what type of action? Listen to the debate and you will hear many different kinds of arguments – about whether and how the climate is changing, whether human activities are responsible, how much of the change occurring might be natural, how the climate might change in the future, what the effects of the changes will be and whether they matter, and the feasibility, advantages, and disadvantages of various responses. Although these arguments are distinct, when advanced in policy debate they all serve to make a case for what we should or should not do. They aim to convince others to support a particular course of action.

This chapter lays the foundation for understanding these arguments. The next section lays out the differences between the two kinds of claims advanced in policy debates, positive and normative claims. Sections 2.2 and 2.3 then discuss how science examines and tests positive claims, and how participants in policy debates use both positive and normative claims to build arguments for and against proposed courses of action. Section 2.4 examines what happens when scientific and policy debates intersect, as they do for climate change. Finally, Section 2.5 discusses the role of scientific assessment in managing the boundary between scientific and policy debate. Later chapters discuss the specific claims people advance about the science and policy of climate change, and the state of present knowledge on these claims.

2.1 Justifications for action: positive and normative statements

On climate change, as on any issue where people disagree about what to do, the arguments advanced to support or oppose proposed action rest on two fundamentally different kinds of supporting statements: statements about what we know, or *positive claims*, and statements about what we value or should value, or *normative claims*.

A positive claim concerns the way things are: it says that something is true about the world. It might concern some state of affairs ("It is raining"), a trend over time ("Winters are getting warmer"), or a causal relationship that explains why something happens ("Smoking causes cancer"). Positive statements do not have to be simple or easy to verify, and they may concern human affairs as well as the biophysical world. "US foreign policy during the Cold War contributed decisively to the collapse of the Soviet Union" is also a positive statement, although one that would be hard to verify with confidence. What is essential to positive claims is that they concern how things are, not how they should be. All scientific claims and all scientific questions are positive.

A normative claim concerns evaluation: not how things are, but how they should be. It says that something is good or bad, right or wrong, desirable or undesirable, just or unjust, and so on. Examples of normative statements would include, "He should have stayed to help her," "Killing is wrong," "Present inequity in world wealth is unjust," "We have an obligation to protect the Earth," and "Environmental regulations are an unacceptable infringement on property rights and individual liberties." With few exceptions, statements or questions that include the words "should" or "ought" are normative. And the exceptions mostly involve sloppy use of language. If someone says, "The Yankees should win the World Series," he probably means they are *likely* to win (a positive claim), not that it is right or just or proper that they win (a normative claim). Of course, he might mean both these things, providing an example of how we sometimes combine – and confuse – positive and normative statements.

There are several important differences between positive claims or questions, and normative ones. First, if a positive question is sufficiently well posed – meaning all the terms in it are defined clearly and precisely enough – it has right and wrong answers. Similarly, a well-posed positive claim is either true or false. Second, the answer to a positive question, or the truth or falsity of a positive claim, does not depend on who you are, what you like or value, your culture, your political ideology, or your religious beliefs. Finally, arguments over positive claims can often be resolved by looking at evidence. If you and I disagree over whether it is raining, we can look outside. If we disagree over whether winters are getting warmer, we can look at the records of past and present winter temperatures. If we disagree over whether smoking causes cancer, we can look at the health records of a large group of smokers and non-smokers (who are otherwise similar), and observe whether more of the smokers get cancer.

But notice the sneaky word "often" that qualifies our statement above that positive disagreements can be resolved by looking at evidence. Looking at evidence cannot always resolve positive disagreements for two reasons, one philosophical and one practical. Philosophically, there is no rock-solid foundation for authoritatively resolving even positive questions, because you and I might disagree over what the evidence means. We might disagree over the validity of the methods used to compare winter temperatures in different places or over time. We might even disagree about whether what is happening outside right now counts as "rain." (Does a faint drizzle count? What about a thick fog?) If we are stuck in disagreement over such questions of evidence, neither of us can authoritatively win the argument. The best I can do is resort to secondary arguments, like what it is reasonable to believe, or whose judgment to trust, which you might also reject.

The second, practical limitation is that the evidence needed to resolve a disagreement might be presently unavailable, or even unobtainable in principle. We cannot tell whether winters are getting warmer unless we have temperature records over the region and the time period we are asking about. But while these limitations are real, they do not negate the broad generalization: looking at evidence provides a powerful and frequently effective way to resolve disagreements over positive claims.

This is not so for normative claims. Because normative questions always involve value judgments, the basis for believing that they have right and wrong answers is much weaker than for positive questions. Specific normative claims need to be based on some underlying set of principles that define the relevant values. These might be a set of religious beliefs, a moral philosophy, a political ideology, a set of cultural norms, or they might simply refer to people's preferences or interests (what people want, or what is good for them). But because people have deep differences over such underlying principles, the answer to a normative question can differ widely from person to person. Even a claim like "killing is wrong," which might at first seem obviously true, quickly generates disagreement when you consider hard cases such as euthanasia, capital punishment, or war. Moreover, looking at evidence is of no help in resolving differences over purely normative questions. Normative questions are consequently more deeply contested than positive ones, and less amenable to mutually agreed resolution.

In policy debates, the arguments for a particular course of action nearly always rest on both positive and normative claims. This is because most policy choices are made for instrumental reasons: we advocate doing something because we think it is likely to bring about good consequences. Arguments about actions ("Shall we raise the tax on cigarettes?") then depend partly on positive arguments about what their consequences will be ("If we raise the tax, how much less will people smoke?", "What health benefits will this reduction in smoking bring?", "How much revenue will be raised, and from whom?", "How much cigarette smuggling will there be?"). They also depend on normative arguments about how good or bad these consequences are ("Is it fair to raise tax revenues from the poor?", "Is it worth accepting the projected increase in crime to gain the projected health benefits?"); and also on normative arguments about the acceptability of the action itself ("Is trying to make people reduce unhealthy behavior the proper business of the government?"). Similarly, people in favor of capital punishment argue that it deters people from committing heinous crimes (positive), that its application is not racially biased (positive), that procedural safeguards can reduce the risk of executing the innocent to nearly zero (positive), that murderers deserve to die (normative), and that it is just and legitimate for the state to execute them (normative). Opponents argue that deterrence is ineffective (positive), that sentencing outcomes are racially biased (positive), that the rate of errors – executing innocent people – is and will remain high (positive), and that it is wrong for the state to kill (normative).

On the climate-change issue, arguments on all sides of the debate also combine positive and normative claims. Proponents of action to reduce greenhouse-gas emissions argue that the climate has warmed, that human actions are largely responsible for recent warming, and that changes are likely to continue and accelerate – all positive claims. They also argue that the resultant impacts on resources, ecosystems, and society are likely to be unacceptably severe, and that we can limit future climate change at acceptable cost – statements that combine positive claims about the character of expected impacts, the opportunities for technological change, and the effectiveness and cost of responses, with normative claims about the acceptability of these costs. All these claims, positive and normative, are disputed by opponents of action to reduce emissions.

But while policy arguments may involve both positive and normative claims, these do not come neatly identified and separately packaged. Rather, many arguments, like those above, intertwine positive and normative elements. For example, consider the statement, "Climate science is too uncertain to justify costly restrictions on our economic growth." This says that restrictions on emissions are not justified, which appears to be a normative claim. But the claim also depends on unstated assumptions about positive matters, including what we know (and how confidently we know it) about how fast the climate is likely to change, what the impacts will be, what means are available to slow the changes, and how costly and difficult these will be. The person making this argument may have considered all these in reaching her judgment that restrictions on emissions are not justified. But hearing this argument, you would have to consider whether she is correct in these assumptions to decide whether or not you agree with her conclusion. You and she might agree completely on what state of scientific knowledge would justify action, but still disagree on the conclusion if you disagree on the state of scientific knowledge.

The unstated assumptions behind an argument can be normative as well as positive. Consider the statement, "The Kyoto Protocol would cost the US economy hundreds of billions of dollars while exempting China and India from any burdens." This says something about the costs of a particular policy, which sounds like a positive claim. But the statement also has rhetorical power, since it strongly implies that it would be wrong or even foolish for the US to join the Kyoto Protocol. Whether or not the positive claim is correct, it gains this rhetorical force from several unstated assumptions, some positive and some normative: that this cost is too high, relative to whatever benefits the Kyoto Protocol might bring the US; that imposing the initial burden of emission reductions on the rich industrialized countries is unfair; and that other courses of action open to the US are better.

This tangling of positive with normative claims, and of explicit arguments with powerful unstated assumptions, obstructs reasoned deliberations on policy decisions. It creates confusion, exacerbates conflict, and makes it difficult for citizens to come to an informed view. This tangling might sometimes be inadvertent, or might be intended to sow confusion in the debate, so as to obscure areas of potential agreement. The pieces of an argument cannot always be perfectly disentangled, of course. But untangling them to the extent possible, and making the major assumptions that underlie policy arguments explicit, can often reduce conflict and identify bases for agreed action among people of diverse political principles.

Separating positive from normative claims is particularly important for environmental issues because of the central role played in these debates by positive claims about the behavior of environmental systems. Participants in environmental policy debates very often present their positions as based on science, even when others are advancing precisely opposing scientific claims. One advocate might say, "Scientific evidence shows that human greenhouse-gas emissions are warming the Earth," while another says, "There is no scientific evidence that human greenhouse-gas emissions are warming the Earth." Assuming the terms in these two statements are defined clearly and consistently, they cannot both be true. Resolving disputes over positive claims can make a substantial contribution to reducing disagreement over what course of action to pursue.

And such resolution is often possible. Indeed, on many environmental issues, relevant knowledge is more advanced and scientific agreement stronger than you would think from reviewing the policy debate in the newspaper or on the web. This is emphatically the case for global climate change. We know more about the climate, how it is changing, and how it is likely to continue changing under continued human pressures, than a look at the policy debate would suggest. To understand why, we first explore how the social process we call "science" works. We then explore how political decision-making works, and what happens when these two very different social processes come into contact with each other.

2.2 How science works

Science is a process that advances our collective knowledge of the world by proposing and testing positive claims. Science is a *social* activity – not in the sense that a party is a social activity, something we do for the purpose of enjoying other people's company, but rather in the sense that a sports team or an orchestra is a social activity: an activity that gains power from harnessing the skills and efforts of multiple people in pursuit of a common goal. The power of the social process of science to answer positive questions and advance our knowledge of the world, while not absolute or perfect, is unparalleled in human history.

As with a sports team or an orchestra, people get to join the community of scientists by training and practicing until they demonstrate that their skills and knowledge are sufficient to contribute to the group objective. Also as with a team or orchestra, there are rules and guidelines that determine how the scientific community pursues its goal and how individual scientists contribute to the collective effort. In science, the rules and guidelines make up the scientific method – a description of what scientists do that appears in the opening pages of every elementary science textbook. Although descriptions of the scientific method differ in detail, at their core all have a three-part logical structure. First, making up proposals or guesses about how the world works – these are called hypotheses. Second, reasoning about what the hypothesis implies for evidence that we should be able to observe. Third, testing the hypothesis by looking at the evidence.

You can use this logical structure of inquiry to investigate any positive question, small or large: "Why do my keys keep disappearing?", "Who killed Cock Robin?", "How do stars form?", "Are people being abducted by aliens?", or "Are human activities warming the Earth?" In particular scientific fields, there are additional constraints on the application of this method that come from the current state of accepted knowledge in the field, which defines what counts as an important question and a plausible, interesting answer. A hypothesis that contradicts well-settled knowledge is regarded – reasonably – as almost sure to be wrong, and so is unlikely to attract any interest. For example, a new proposal that the Earth is fixed in space and heavenly bodies revolve around it, or that microbial infections do not cause disease, would not attract scientific interest.

For a hypothesis to contribute to advancing scientific knowledge, it must be *testable.* This means it must imply specific predictions of things you should be able to observe – beyond those already known, or which suggested the hypothesis – if it is true. It is the specific observable implications of a hypothesis that make it vulnerable to being refuted by evidence. If you look carefully and do not see what the hypothesis says you should see, or see what the hypothesis says you should not, then you conclude the hypothesis is probably wrong. Perhaps it can be adjusted to be consistent with the evidence, but such adding of qualifications and complexity to a hypothesis to account for contrary evidence is regarded with suspicion. A hypothesis that is specific and testable, but when tested is found to be wrong, can still contribute to advancing scientific knowledge. It might, for example, help direct efforts to more fruitful lines of inquiry or stimulate someone to generate a better hypothesis. But a hypothesis with no observable implications, or whose implications are so vague or pliable that it is impossible to say what would count as contrary evidence, is of no use in scientific inquiry. This is why science has nothing to say one way or the other about questions of religious belief, such as the existence of God.

Paternity testing provides a simple illustration of how evidence is used to test a hypothesis. Before DNA testing was developed, known patterns of blood-type inheritance were often used to test who was the biological father of a child when this was disputed. If the mother and child have certain blood types, this limits the possible blood types of the father. For example, if the mother is type A and the child is type B, then the father can only be type B or type AB. Suppose your hypothesis is that James is the father. For this hypothesis to be true, James must have blood type B or AB. If you observe that James has type A blood, then (except for the possibility of an error in the observation or a mixup of samples) this decisively rejects the hypothesis that James is the father. Note, however, that if you find James is type B, the hypothesis that he is the father is not rejected by the evidence, but it is not proven true either. The true father could be James, or could be some other man with type B or AB blood.[1]

This illustrates a general characteristic of scientific inquiry, that hypotheses are rejected more decisively than they are supported. Because hypotheses are constructed to imply certain observable evidence, decisive contrary evidence

[1] Modern genetic testing is more powerful than blood-type testing, because it observes many characteristics. But like blood-type testing, its results are only decisive in *rejecting* a match: if your DNA does not match all the characteristics of the tissue sample, the sample

usually kills the hypothesis; but sometimes supporting evidence can arise by coincidence, even if the hypothesis is wrong. This characteristic is sometimes summarized by saying that science never proves anything, because while a hypothesis that has survived enough repeated testing comes to be accepted as correct, it always remains vulnerable to being disproven by some future test.

In some fields of science, the observations used to test hypotheses are generated through experiments, by isolating the phenomenon of interest in a laboratory and actively manipulating some conditions while controlling others to generate observations that precisely target the hypothesis to be tested. You can do this if you are studying chemical reactions, or the behavior of semiconductors, or the genetics of fruit flies. But for some scientific questions, such as questions about the behavior of the Earth's atmosphere, the formation of stars, or evolution of life in the distant past, you cannot do such controlled experiments in a laboratory. It is not possible, nor would it be acceptable, to put the Earth in a laboratory and manipulate some characteristic of the atmosphere to observe the response. But it is often still possible to observe naturally occurring processes in order to piece together the evidence needed to test the hypothesis.

For example, Einstein's theory of general relativity says that gravity should bend the path of a beam of light, just as it bends the path of a ball thrown into the air. The astronomer Sir Arthur Eddington saw that this part of the theory could be tested by observing the position of a group of stars when their location, as seen from Earth, lies very close to the edge of the Sun. If light traveling from a star to the Earth bends as it passes through the strong gravitational field close to the Sun, then the star's position (measured relative to other stars) should appear to be shifted from its location when observed in the night sky. The Sun is so bright, however, that the only way to observe a star's apparent location when it is near the Sun is during a solar eclipse. Eddington's group traveled to Principe, off the coast of Africa, to photograph stars during an eclipse on May 29, 1919. Comparing these photographs to photos of the same stars at night showed that the light had indeed been bent by the Sun's gravitational pull, by an amount that was close to what the theory of general relativity predicted.

The work done by individual scientists or teams is only the first step in the social process of science. Whether the work proposes a theoretical claim ("I have a new explanation for the ozone hole") or an observation ("I have a new measurement of the flow of carbon between forests and the atmosphere"), it

did not come from you. If you do match all the characteristics, the sample probably came from you but this is not certain. In one early form of DNA paternity testing, for example, a perfect match still left roughly a 0.2 percent chance – 2 chances in a thousand – that the father is not you, but someone else who matched all the tested characteristics.

must then be judged by the relevant scientific community. This process sta with writing up the work and results – with a description of exactly what w done and how, the data, and the calculations or other methods of analysis, i ally in enough detail that someone knowledgeable in the field could reproduce the work – and submitting it for publication in a scientific journal.

The first formal control that the scientific community exercises on the quality of scientific work comes at this point. Scientific journals will not publish a paper until it has been critically examined by other scientists who are experts on its subject. In this process, called peer review, the reviewers' job is to look for any errors or weaknesses – in data used, calculations, experimental methods, or interpretation of results – that might cast doubt on the paper's conclusions. The process is usually anonymous, so reviewers are free to give their honest professional opinion without fear of embarrassment or retribution.

Succeeding at peer review counts for everything in a scientific career. For scientific work to attract attention and respect, it has to be published in peer-reviewed journals. Proposals for research funding must also go through peer review. For scientists to get and keep jobs and achieve all other forms of professional reward and status, they must succeed at getting their work through peer review.

Peer review is a highly effective filter, which stops most errors from being published, but it cannot catch every problem. Reviewers occasionally fail to notice an obvious mistake, and there are some types of error that reviewers cannot catch. They cannot tell if the author misread observations of an instrument, or wrote a number down wrong, or if chemical samples used in an experiment were contaminated. Moreover, peer review often cannot identify clever fraud, such as the rare cases where the scientific work being reported was not really done at all.

But peer review is only the first of many levels of testing and quality control applied to scientific claims. When an important or novel claim is published in a journal, other scientists test the result by trying to replicate it, often using different data sets, experimental designs, or analytic techniques. While one scientist might make a mistake, do a sloppy experiment, or misinterpret their results (and peer reviewers might fail to catch it), it is unlikely that several independent groups will make the same mistake. Consequently, as other scientists repeat an observation, or examine a question using different approaches and get the same answer, the community increasingly comes to accept the claim as correct.

For example, early in the ozone depletion controversy, theory suggested that releases of chlorofluorocarbons (CFCs) should be causing a reduction in ozone, but no reduction could be seen in the observations available at the time. In the early 1980s, a few scientists began proposing that a decline could be observed in the latest ozone measurements. There were many problems with the data,

How tough is peer review really?

Very tough. You might expect peer review to be a rubber stamp, or a way for scientists to pat each other on the back, but it is usually a highly critical examination of work proposed for publication. The following rejection letter from a journal editor (edited for anonymity) gives a taste of how demanding the process is.

Dear Dr. Smith,

I have received the reviews of your paper, "Isotopes, seasonal signals, and transport near the tropical tropopause". On the basis of these reviews, I regret that I cannot accept the paper for publication in its present form. This was a difficult decision, since Reviewers A and B recommend rejection, while Reviewer C is more positive about the study. Yet even Reviewer C has serious misgivings, about potential numerical problems in the model and insufficient comparison of the results to observations. For their part, Reviewers A and B are thoroughly unconvinced that the model is sufficiently constrained by the few observations available. All reviewers are also concerned that the model's sensitivity to many tunable parameters makes the results suspect. Given the seriousness of these issues, I cannot accept the manuscript. However, since Reviewer A has suggested the study could be reworked to be acceptable and Reviewer C is generally supportive, I encourage you to revise the paper thoroughly and resubmit a new version – if, that is, you think the concerns can be dealt with adequately. In that regard, Reviewer A argues for a more thorough sensitivity analysis, and all reviewers call for detailed justification of the many decisions made in tuning the model. This should be done with reference to observations wherever possible, but where not, physical arguments and results from previous studies may also be used. If you choose this course, please pay careful attention to all comments of the reviewers, major and minor, and provide a detailed, point-by-point response to each reviewer.

Regards,
John Q. Pseudonym, Editor

What does this mean? Reviewers A and B were not convinced the paper's scientific analysis supported its conclusions. Although reviewer C recommended accepting the paper, the editor looked carefully at the reviews and the paper, decided he agreed with reviewers A and B, and rejected the paper. But while this version is not acceptable, the authors might still succeed at making the work publishable. The editor advises them to revise the work, address the reviewers' criticisms, and try again.

however, and when other scientists examined them, they concluded the reductions being proposed could not be distinguished from deterioration of the measuring instrument, which was known to be occurring. As a result, the claims were rejected. Then in 1988, a new analysis including more recent data suggested stronger evidence of a decline. Because this claim was so important, three other scientific teams checked and re-analyzed the data behind this new claim, as well as analyzing related data. This time the other teams also found a decrease in ozone, similar in size to that calculated by the first team. The conclusion was therefore confirmed, and atmospheric scientists accepted that there now was a real decline in global ozone.

This multi-layered process of criticizing, testing, and replicating new scientific claims is public, collective, and impersonal. Individual scientists make mistakes, and are prone to biases, enthusiasms, or ambitions that may cloud their vision, as we all are. But however intensely a scientist may hope for honor from having his novel claim accepted, or want a result consistent with his political beliefs or financial interests, scientists know that any claim they propose, especially if it is important, will be critically examined by other scientists and sloppy, biased, or weakly supported work is likely to be exposed. Moreover, scientists confer respect and status on those who are careful in their work, critical and fair in their argument, and cautious in advancing claims. Intemperate claims, partisan or biased testing, or less than scrupulously honest reporting of results can damage a reputation so severely that scientists have strong incentives to be cautious.

The result of this process of collective testing, and the incentives embedded in it, is to make science highly conservative. The burden of proof lies with the person making any claim that extends present knowledge or contradicts present belief. The more important and novel the claim being advanced, the more aggressive the scrutiny and testing it will receive and the higher the standard of evidence required to accept it: remarkable claims require remarkable evidence. This is the way science maintains the stability of the received body of knowledge and protects against errors and fads.

This process of making hypotheses and testing them through careful, repeated observations does not generate proven truth. Science never *proves* anything. Even a hypothesis that has survived repeated testing and come to be accepted remains vulnerable to being overturned by some future test. But some claims are so well verified, by accumulating independent evidence, resolving controversies, and rejecting contrary claims, that they come to be regarded simply as facts. For example, we now accept as facts that the structure of DNA is a double-helix, that atoms obey the laws of quantum mechanics, and that burning fossil fuels has increased the abundance of CO_2 in the atmosphere. These claims have

Is this really how science works?

Not exactly, but close enough. This description is a simplification of how science is actually practiced. Several decades of research in the history and sociology of science has fleshed out how and how much the actual practice of science diverges from this model, in particular how social factors impinge on the practice of science. The most basic insight was that of Thomas Kuhn, who recognized that normal progress in a scientific field depends upon a deep level of shared assumptions that define what questions are important, what lines of inquiry are promising, and what hypotheses are plausible and interesting. These shared presumptions, which Kuhn called "paradigms," are not explicitly examined or even necessarily recognized by the scientists who hold them. Paradigms change only infrequently, in revolutionary periods that follow the accumulation of some critical mass of "anomalies" – results that don't fit the accepted model, but are provisionally set aside.

Science is not an abstract, rational process, but a collective human endeavor. As such, its actual practice diverges from this idealized description in various ways. Social factors such as status, charisma, and rhetorical skill do to some extent influence whose arguments get paid attention and are trusted. Not all claims are immediately tested and replicated. And consensus views of what questions are interesting and important are not formed on purely rational bases. But the power of these social processes to influence the content of what comes to be accepted as scientific knowledge is limited and provisional. Conspicuous claims that do not stand up to testing eventually get rejected, no matter who is supporting them. Accepted beliefs that accumulate enough anomalies are eventually re-examined, revised, or rejected, however comfortable or fashionable they may be.

been so well verified that further testing of them is considered unnecessary and uninteresting.

How can you tell whether the relevant scientific community has accepted a claim as "true"? The most reliable way is to look in the peer-reviewed literature for multiple, independent verifications. Other factors count in addition to the number of verifications, of course. Some tests are more stringent than others, for example, just as DNA paternity testing is more stringent – more likely to reject a match – than blood-type testing. When newer, well-checked observations contradict older ones, the newer ones are usually given more weight because observational instruments generally improve with advancing technology. The extent to which competing claims have been tested and rejected also matters – if every proposal that has been made but one has been clearly rejected, the remaining one is more likely to be accepted, at least tentatively, even if the affirmative evidence supporting it is not conclusive.

The reputations of the scientists involved also affect the community's willingness to believe a claim. The same work is likely to be given more credence when done by a scientist with a well-established reputation for careful, competent work than when done by one who is unknown or known to have done sloppy work in the past. Eddington's confirmation of general relativity carried more weight, and was probably accepted more quickly, because of his preeminent reputation for scrupulously careful observations.

Even though scientific knowledge is always provisional and never proven, a strong scientific consensus provides a better basis for relying on the truth of a positive claim than is provided by any other human process for pursuing knowledge. It is of course possible for a strong scientific consensus to be wrong. We are reminded of this possibility on those occasions when new results contradict and eventually overturn a previously accepted understanding. But while this does happen, and generates much excitement and attention when it does, it occurs infrequently.

The risk of a consensus later being found to be wrong is greater for some types of scientific claims than for others. The risk is greatest for fundamental theories, particularly if their strongest predictions concern matters that are beyond our present capability to observe. We would not be too surprised if Einstein's theory of gravitation were one day superseded by another, just as Newton's theory of gravitation was previously superseded by Einstein's. The risk is smallest for simple, concrete claims, such as a single observation or measurement. When an observation has been repeatedly checked using various methods and accepted as correct, it is quite unlikely to be subsequently overturned. Between these extremes, claims about the associations between different observations from which we infer cause and effect are on somewhat weaker, although still very strong, ground. We would be extremely surprised to learn in the future that smoking does not increase the risk of cancer, or that chlorine chemistry does not cause the Antarctic ozone hole, although in principle either of these could happen.

The lesson we draw from this discussion is that when there is a strong scientific consensus on some positive point, those outside the relevant scientific community who care about the answer can do no better than relying on the consensus. Unfortunately, this advice is not always helpful, for two reasons: a

An example: the discovery and explanation of the Antarctic ozone hole

The discovery and verification of the Antarctic ozone hole and the search for its cause illustrate how scientific claims are checked and hypotheses are tested. In 1982, researchers with the British Antarctic Survey noticed that the total amount of ozone over their station in October – early spring in Antarctica – appeared to be dropping sharply from levels of the 1960s and 1970s. (Figure 2.1 shows their data, extended to the mid-1990s.) Ozone in other months appeared normal. At first concerned they might have an instrument error – always a likely explanation for a wildly unexpected observation – the scientists spent two years checking and confirming their results before submitting them to the scientific journal Nature, where their paper appeared in June 1985. The observations created a scientific firestorm. The results had been peer-reviewed, but such a dramatic result demanded independent verification. This was quickly obtained by reviewing archived data from a satellite instrument, and by further measurements by several groups.

So the losses were real, but what was causing them? Atmospheric scientists had been predicting for ten years that we should see ozone depletion from CFCs. But these observed losses were occurring at a time and place virtually opposite to where the CFC-ozone theory said they should be, and were also much larger. It was therefore not clear whether CFCs were the cause, or something else.

Over the following year, three competing theories were proposed to explain the shocking losses, each implying different things that should be observable in the region of ozone loss. Observations made in 1986 and 1987 decisively rejected two of these theories, and provided compelling support for the third.

The first theory proposed that ozone was being destroyed by naturally occurring nitrogen oxides, which increased in the Antarctic stratosphere after the 1980 peak of the 11-year cycle of solar energy. If this was true, several pieces of evidence should have been present. Ozone loss should be accompanied by high concentrations of nitrogen oxides, and should be greatest in the upper stratosphere where most nitrogen oxide production occurs. Moreover, there should have been similar ozone losses after prior solar-maximum years such as 1958 and 1969, and the losses should begin to reverse by the mid-1980s. Observations in 1986 found none of these predictions was confirmed, quickly ruling out this hypothesis: ozone losses were largest in the lower stratosphere and were accompanied by

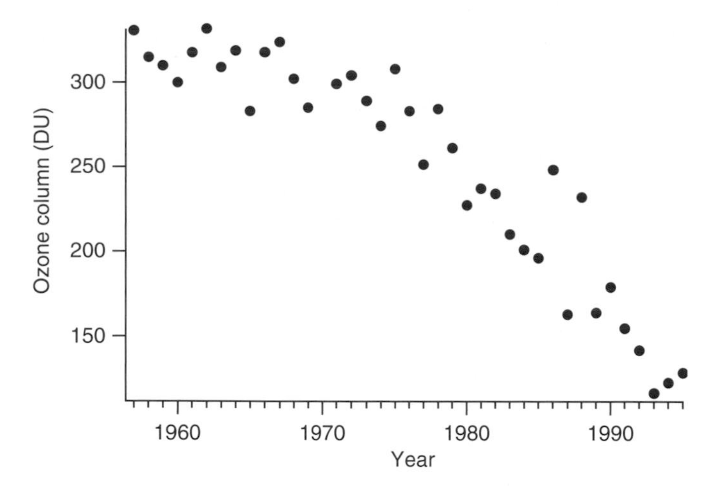

Figure 2.1 Total atmospheric ozone in October over Halley Bay, Antarctica, measured by the British Antarctic Survey. The amount of column ozone is expressed in Dobson Units (DU).
Source: adapted from Fig. 7.2 of Dessler [2000].

low, not high, nitrogen oxides; moreover, ozone records showed no similar losses after earlier solar maxima, and losses were clearly accelerating, not reversing, through the 1980s.

The second theory proposed that changes in worldwide movement of stratospheric air were reducing the transport of ozone to Antarctica. This theory also predicted several specific observations. Temperatures in the loss region should be unusually cold. The vortex of stratospheric winds that surrounds Antarctica in winter should be stronger than usual and persist longer into the spring. And most crucially, the general movement of air below and within the loss region should be upward. Although initial observations gave limited support to this theory – the vortex did appear to be unusually strong, and Octobers appeared to have grown colder – this theory was rejected by 1987 when it became clear that Antarctic air was generally sinking, not rising as the theory required.

The third theory proposed that ozone was being destroyed by newly identified chemical reactions catalyzed by chlorine carried to the stratosphere by CFCs. Several groups of scientists proposed different reactions, but these all required that a particular chlorine species, chlorine monoxide (ClO), should be abundant where ozone loss was occurring. Simultaneous observations of ozone and ClO from flights through the ozone hole in September 1987 found remarkably strong support for this hypothesis. ClO suddenly increased a hundred-fold each time the aircraft entered the region of ozone loss, and dropped again as it left. This negative correlation between ClO and ozone – whenever ClO went up, ozone went down – was so exact that many observers called this result the "smoking gun" – one of those rare instances where a hypothesis comes to be accepted on the basis of a single, compelling observation. Over the next few months, as the results were reviewed for publication and discussed at scientific meetings, there formed a strong consensus, which has been sustained and elaborated since, that CFCs and related chemicals were the principal cause of the ozone hole.

strong consensus on a policy-relevant scientific question might not exist; or a consensus might exist but be difficult for anyone outside the field to observe. A consensus might not exist because a key question lies beyond present knowledge or research capabilities, or has simply not attracted much scientific effort – perhaps because, despite its perceived importance to policy, the question is regarded as low in scientific interest. Alternatively, a question might be under investigation, accumulating evidence but not fully resolved, perhaps with scientists disagreeing over how well settled it is. One implication of the cautious and conservative nature of science is that acceptance of new claims happens slowly, often much more slowly than answers are demanded by the policy-making process.

Alternatively, a consensus might be present in a scientific community but hard to see for anyone outside the field. Scientific discussions focus on what is interesting to scientists: not what is well established, but what is new, uncertain, and controversial. Moreover, even when a strong consensus exists on some point, that may be obscured for those outside the field by a few vocal advocates of an opposing view, even if that opposing view has been decisively rejected. Since even scientists in other fields may lack the specific knowledge to judge the merits of opposing claims in the specialist literature, non-scientific policy actors cannot hope to make such independent judgments themselves. Rather, they must rely on some form of summary and synthesis of what the scientific community knows and how confidently it knows it. Section 2.5 discusses the role of scientific assessment bodies in providing such a synthesis.

2.3 Politics and policy debates

Politics is concerned not with positive questions, but with collective action: not what is true, but what shall we do? Politics embraces the processes of argument, negotiation, and struggle over joint actions or decisions – most often the decisions of what policies will be adopted by government institutions. Like most competitive arenas, politics involves conflict but the conflict is bounded. In competitive sports, the boundaries are defined by the rules of the game as enforced and interpreted by the referees. In politics, the boundaries are defined by the structure of rules and institutions within which decisions are made. In any nation, policy-making takes place within a complex structure of constitutions, laws, and traditions that grants specific authorities to make decisions and imposes various constraints on the exercise of that authority.

Sometimes the authority to make decisions is simple and absolute: the President of the United States has the power to grant pardons for federal criminal offences. More often, however, authority is limited by rules that restrict how it can be exercised, or by related authority held by others. Under the US Clean Air Act, for example, the Environmental Protection Agency (EPA) has the authority to enact regulations restricting chemicals that deplete the ozone layer. But to do this, the EPA must present evidence that the chemical being restricted is a strong enough ozone depleter to fall under the requirements of the law, must publicize proposed regulations for a 90-day comment period, and must respond to comments received before enacting final regulations. If the EPA fails to meet these requirements, parties who disagree with the regulation might be able to have it overturned by challenging it in court. Moreover, since the EPA's authority to make these regulations is delegated to them by Congress in the Clean Air Act, Congress could revise or revoke the authority by amending the Act.[2]

Sometimes, particularly for issues on which the government has not previously acted, the relevant authority might be widely distributed, defined only vaguely, or not defined at all. The more novel the issue and the less clear the existing lines of authority, the more fluid is the political process in terms of what the key decisions are, who makes them, who gets to influence them, and what factors contribute to the outcome. Different countries' political systems differ in their overall openness to public influence over decisions, and in their particular channels for influence. Even in highly open systems like the United States, however, exercising influence over a policy decision is difficult, requiring much time, energy, money, strategic skill, or luck. Because it is so difficult,

[2] Since this section of the law meets a US obligation under an international treaty, the Montreal Protocol, changing the law to revoke this authority would put the United States in violation of the treaty. But while this might make Congress think twice about making such a change, it does not eliminate their authority to do it, since the United States has the authority to withdraw from international treaties.

those who do mobilize to influence policy on any issue normally represent a tiny fraction of the electorate.[3]

The few people who do mobilize to influence policy do so for many reasons. Some may hold strong views about the right thing to do or the best interests of the nation. Some may expect proposed decisions to benefit or harm them in some concrete way – e.g., affecting the health or well-being of their families, helping or hurting their livelihoods, or affecting the value of their property or the profitability of their business. Some may have ambitions to exercise political power or influence. Any of these motivations can bring conflicts between groups seeking to influence policy. My group might compete with yours to be a key deal-maker on environmental policy; a proposed regulation might benefit my industry but harm yours; or I might passionately believe that the environment needs further protection, while you believe with equal passion and sincerity that environmental regulations threaten basic liberties.

People sometimes speak, as we do here, about the "policy debate" on public issues like climate change. But describing policy-making as a debate is somewhat misleading. You win a debate by persuading people – impartial third parties, and sometimes your opponents – that the arguments favoring your course of action are stronger than those on the other side. In policy-making, the strength of arguments on each side does matter, but as one factor among many that influence what happens. Policy actors use many methods to build support for the decisions they want: well-founded rational arguments when these are available, but also biased or inaccurate arguments, appeals to emotion or prejudice, flattery and manipulation, promises and threats, deals to exchange support on other issues, and sometimes – although these are illegal in most nations – bribery and coercion.

But in this complex mix of factors shaping policy decisions, rational arguments usually matter at least a little, and sometimes a lot. Arguments matter more when an issue is prominent enough, and is perceived to have high enough stakes, that it attracts public attention and media scrutiny. Such scrutiny increases legislators' and officials' concern with acting competently and

[3] That few people get active on an issue does not, however, mean that few care about it; public opinion comes in all shapes and sizes. No issue gets onto the political agenda without at least a few people caring strongly about it. But the rest of the public might or might not understand or care about the issue, and those who do care might be largely in agreement or strongly divided. Those who are active on an issue try to claim their position is strongly supported by a "silent majority" when they can. Public opinions tend to be strongest and most divided on issues that raise deep differences of principle, such as abortion or capital punishment. While most people express strong support for environmental protection, this only infrequently rises to the intensity of opinion typical of more morally charged issues.

impartially in the public interest – and being seen to act so – and so reduces the scope for more crass and sneaky forms of political influence that would be embarrassing if exposed. Rational arguments also matter more when an issue is so novel that its character, relevant analogies to other issues, and the consequences of alternative actions, are unclear. Under these conditions, many actors may be uncertain what course to favor: their general political principles may give little guidance, and their interests may be unclear. Climate change has these characteristics: enough salience that political decision-making is subject to heightened scrutiny, enough novelty to challenge existing lines of authority, and enough uncertainty that many actors do not line up predictably according to either their general political principles or how the issue is going to affect them – although over the past few years, the issue has grown significantly more partisan in the United States.

Consequently, climate change is an issue on which we expect rational arguments, both positive and normative, to be influential. On normative arguments, an explicit debate between competing perspectives, some of which may be directly in contradiction, makes sense. If one group in the climate change debate argues that our primary obligation is to protect the environment, while another argues for the primacy of individual freedoms, neither of these is right or wrong. It is entirely appropriate for proponents of these contending views to compete to persuade policy-makers and citizens. Science and scientists have no special authority in this debate. On relevant positive arguments, however, such as observed changes in the climate, the causes of these changes, and the likely nature of future climate change under additional human inputs, scientific knowledge does have special authority and therefore much to contribute. But this potential contribution is often obstructed by a lack of understanding of the differences between scientific arguments and policy arguments.

The most important of these differences concern the motivations of participants, and the rules under which they operate. Scientists gain professional status by advancing shared knowledge, but also by being cautious in interpreting new claims and fair in their assessment of competing claims. While there are real disagreements and rivalries in scientific debates, these norms introduce substantial elements of shared interest, so scientific disputes are rarely pure, "zero-sum" conflicts. Advancing knowledge benefits the scientific enterprise as a whole, even if the person making the discovery benefits the most.

In policy debates, there can also be widely shared interests – e.g., in not wasting public money on worthless projects or corruption, or protecting the nation against threats from hostile foreign powers – but competing interests are more prominent. Political actors are rewarded for succeeding in various ways where one's gain is another's loss, such as gaining and holding power, enacting policies consistent with their political principles, and delivering the benefits of government action (e.g., spending on public-works projects) to their supporters and constituents. Even when an issue like climate change is so novel and uncertain that people see less clearly where their material interests lie, these incentives still introduce competitive elements into all policy decisions.

A second basic difference between scientific and political arguments concerns the rules of acceptable argument. The rules in both domains are mostly unwritten, enforced only by the approval and censure of others, but they still matter. The rules of scientific argument are highly constraining. Whatever their true motivation, scientists must argue as if motivated purely by the pursuit of knowledge. A scientist who breaks the rules – who makes sweeping claims based on limited evidence, fails to acknowledge how he could be wrong, selects evidence opportunistically to support his view, ignores or misrepresents contrary evidence, makes emotion-laden arguments, or makes personal attacks on opponents – risks irreparable harm to his reputation and professional standing.

The rules of policy argument are much more lenient. In policy debates, exaggerated, selective, or biased claims, appeals to emotion, and personal attacks of some minimal relevance to the matter at issue, are often effective and rarely punished or even censured. Even more aggressive tactics, such as personal attacks of no substantive relevance, appeals to prejudice, and outright lies, are only weakly restrained. Public outrage might do so, but rarely lasts long enough to be effective, while political opponents must be careful in calling such tactics to account, since they may sometimes use them too. Moreover, there are many ways to gain standing in a policy debate. Having a reputation for knowledge and honesty is one way, but so is representing an important constituency, or the self-fulfilling perception that you are likely to be influential. Consequently, losing scientific credibility does not necessarily jeopardize standing in policy debates. In view of the lesser consequences from going over the line, policy advocates are more willing than scientists to take risks with their credibility to advance their objectives.

2.4 When science and politics meet

Many policy debates, including virtually all environmental issues, depend in part on positive claims about the world that scientific knowledge, in principle, could resolve. Whether we want to spend money and impose restrictions on some activity depends in part on what environmental changes it is causing and is likely to cause – their character, size, speed, and their effects on people and things they care about. These are positive questions, potentially answerable by science.

But science alone – or for that matter, positive claims alone – cannot tell us what to do. Even if the environmental consequences of alternative actions are fully and confidently described, deciding what to do still requires evaluating how much we care about the consequences and what burdens we are willing to accept to reduce them. Nevertheless, scientific claims can be powerful proxies for preferred action. This is easiest to see in extreme cases. If scientific evidence confidently indicated that some technology or activity posed no environmental risk, it is clear that there would be no support for restricting the activity on environmental grounds. Conversely, if compelling evidence showed some activity was likely to bring changes so catastrophic they threatened to end human life on Earth, we would not have to worry much about whether people judged this a bad thing before deciding to restrict the activity.

We rarely have such high confidence about extreme outcomes, of course. But even in real policy debates, there is widespread support for some degree of precaution on environmental issues – avoiding serious environmental changes that are sufficiently well established, even if this means bearing some costs and burdens. This gives special power to scientific claims about how human activities change the environment, despite the room for political disagreement buried in those qualifiers in the previous sentence: "serious changes" (partly a positive question of what changes, partly a normative question of how to evaluate them); "sufficiently well established" (partly a positive question of how confidently established, partly a normative question of what confidence would justify action); and "some costs and burdens" (partly a positive question of what costs and burdens are required to reduce the risk a specified degree, partly a normative question of what costs are warranted to achieve such a reduction).

This special status of scientific claims means that scientific debates and environmental policy debates are closely linked, but this linkage appears in two distinct ways. On the one hand, scientific knowledge about the consequences of alternative courses of action is necessary for responsible public decision-making on environmental issues. On the other hand, every political actor wants to claim their positions are based on science, no matter what their position is. This is partly a general rhetorical tactic, since any position that can take on science's reputation for disinterested pursuit of truth will appear more persuasive. But it also serves the narrower purpose of advancing specific positions. If you can persuade people that scientific evidence indicates an environmental risk is well established and grave, they are more likely to support costly action or restrictions to avoid the risk; persuade them of the contrary, and they are more likely to reject the restrictions. This is why so many political actors, many of them relatively ignorant of science, pretend to engage in scientific arguments.

Because of the different goals and rules of the scientific and political domains, this messy – but unavoidable – interaction between science and politics poses challenges to both sides. Politics challenges science because the positive questions that come to be identified as relevant for policy decisions may be badly posed, or framed by political actors for rhetorical purposes to favor their side. For example, questions such as "Is climate change a crisis?" or "Do we face a climate catastrophe?" are likely to frustrate attempts to provide useful scientific input, because they entirely depend on the non-scientific questions of how you define "crisis" or "catastrophe." Even when policy-relevant scientific questions are clearly and fairly posed, they may be beyond present scientific capabilities to answer, now and perhaps for many years. This is the case, for example, for precise predictions of climate change in specific locations: how will the climate change where I live, with what impacts? But political debate often demands clear, confident, fast answers, and is unsympathetic to scientific caution.

Because many scientists prefer to focus on their scientific work, those who choose to enter public debate do so for an odd mixture of motivations. Some are civic-minded and courageous; some believe their scientific status gives special weight to their policy views; and some seek the public forum for its own rewards – the fame or notoriety, sometimes the influence or profit. Moreover, in policy arenas it is difficult to distinguish well established scientific claims from those that are unrepresentative, uninformed, eccentric, or outright dishonest. Even when scientific consensus on a point is strong and well founded, it can be difficult to persuade a lay audience when arguing against a rhetorically skilled opponent, particularly one who operates by the more lenient rules of policy debate.

Individual scientists who wish to contribute responsibly to public debate consequently face a nasty bind. They can try to reflect the state of knowledge and uncertainty responsibly, and risk having their scientific caution be taken as indecisiveness or inability to speak clearly. Or they can set aside their scientific conservatism and try to distill their understanding into simpler terms more likely to resonate and be understood in the public domain, and risk being charged as intemperate or fame-seeking. Some scientists try to resolve this dilemma by drawing clear distinctions between their roles as scientist and as citizen, expressing carefully qualified scientific opinions then explicitly changing their stance to speak more simply and forcefully as a concerned citizen, but this distinction is hard to maintain in practice. A speaker's scientific standing unavoidably provides some additional credence to their policy views, even while using scientific standing to gain a platform in this way puts that standing at some risk. Moreover, those who try to draw this distinction between scientific and personal views are still sometimes censured for politicizing their science, by their scientific colleagues or, more frequently, by policy actors with opposing views. Given all these risks, it is easy to understand why many scientists decline to engage in public debate on issues where their scientific expertise is relevant, even though doing so is arguably their civic responsibility.

The interaction between scientific and policy arguments also poses hard challenges to non-scientific policy actors, because with rare exceptions, they lack the time and training to critically read the peer-reviewed literature. Attempts by non-specialists to independently evaluate scientific claims carry a large risk of error, because even completely spurious claims can be made to appear plausible to someone who does not know the field. Once again, the history of the ozone-layer debate provides a vivid example. In 1974, within months of the first suggestion that chlorine from chlorofluorocarbons (CFCs) could destroy stratospheric ozone, political opponents of CFC restrictions began claiming that this was impossible because CFC molecules are so much heavier than air that they could never rise to the stratosphere. This sounds like common sense: nearly everyone has seen a mixture of fluids of different densities, like gasoline in water or oil and vinegar in salad dressing, in which the heavier ones settle to the bottom. But common sense or not, this claim is obviously false to anyone who knows how the atmosphere behaves. The atmosphere is not a quiet isolated vessel, but is continuously stirred by winds, vertically as well as horizontally, so long-lived gases like CFCs are mixed to uniform concentrations from the surface to well above the stratosphere, regardless of their weight. Despite being obviously false to anyone with the relevant knowledge, this claim has repeatedly reappeared in the public debate for more than 30 years, as supposed evidence that CFCs cannot harm the ozone layer.

The difficulty of evaluating contending scientific claims is not a problem for policy actors who simply want to select scientific claims to support their positions. The lenient standards of policy debate give these actors a great deal of freedom. They can scan the large number of scientific papers published on a particular controversy to find a few that support their case, even if these are old, known to be erroneous, or decisively refuted by other work. Of the thousands of people with scientific credentials, they can usually find a few who are contrarian or opportunistic enough to go on the record making claims that virtually everyone working in the field knows to be false. As recently as a few years ago, there were still a few scientists willing to state publicly that there was no scientifically persuasive evidence of a link between smoking and cancer. Advocates with enough resources can finance such individuals' participation in policy debates, or even fund programs of research they think likely to generate favorable results.

To succeed at this strategy, it is not necessary to win specific arguments. The status quo enjoys a large advantage in any policy debate, in that it takes substantial political energy to make any change. On environmental issues, the status quo is usually that the activity or technology of concern is not restricted. Consequently, opponents of additional restrictions may succeed by simply exaggerating scientific dissent, if they persuade people that the scientific knowledge indicating additional risk is "too uncertain" to justify a change from the status quo. This can sometimes be achieved merely by advancing enough arguments, even if they are all bad ones, to confuse people. With one claim seeming to cancel an opposing claim, even a settled argument can be made to look like a draw, strongly favoring the status quo.

With these tactics widespread, environmental policy debates contain a cacophony of contending scientific claims and counter-claims, which are circulated without regard for the strength of their evidence or the numbers and stature of the scientists supporting them. Well-established claims backed by near-universal scientific consensus cannot readily be distinguished from the unsupported views of tiny partisan minorities. Many citizens and policy-makers consequently perceive rampant ignorance and uncertainty even where much is well known, and perceive serious controversy even where there is overwhelming consensus.

Under these conditions, the options for citizens or policy actors to inform themselves are limited. Some may simply choose to believe the claims that are consistent with their policy preferences, or side with others who share their general political views – but this approach is dangerous, since there is no reason to expect that nature will conveniently align with your preferences or political views. Some may simply withdraw from the issue and leave it to others to fight it out. Those who want to consider the actual state of scientific knowledge in deciding what actions to support can have difficulty telling what that is. For most, there is no alternative to relying on some level of trust, either in individuals or in institutions. But it is hard to know whom or what to trust, and how much to trust them. Some may have a trusted advisor with relevant scientific expertise, but on any given issue or controversy most will not.

The press is often of little help. Journalists frequently do not understand scientific issues any better than policy actors. Even when they do, journalists follow a professional norm of providing balance between opposing views. Moreover, controversy sells newspapers. Since even settled issues may be debated by a minority, the press generally over-reports scientific dissent and under-reports consensus. Worse still, coverage often favors the dramatic, so the press may give particular prominence not just to minority views, but to extreme views. One scientist's speculation that global climate change may trigger a sudden return to ice-age conditions, or the presentation of such an unfounded scenario in a popular film, makes for dramatic coverage. So do the claims of a half-dozen "climate skeptics" that the scientific consensus on climate change is a political conspiracy. But the careful reporting of the content of that consensus and the evidence supporting it do not.

2.5 Limiting the damage: the role of scientific assessments

The preceding section paints a bleak picture of the pathologies of intertwined scientific and policy debate, but the situation is not hopeless. It is probably unavoidable that climate change is a difficult and contentious issue. But at present, the debate is more confused and contentious than it needs to be, because of widespread misrepresentations of the state of scientific knowledge on relevant positive questions. In our view, it is feasible to structure policy debates so as to reduce the incentives and opportunities for policy actors to practice such misrepresentation.

One essential key to such improvement lies in disentangling the policy debate into separate, clearly posed questions and noting which are positive questions of scientific knowledge about the world, and which are normative judgments about political principles and the evaluation of relevant costs, benefits, and risks. It is not always possible to draw these distinctions cleanly, but trying to do so to the extent feasible can bring large benefits. For individuals engaged in the policy debate, attempting such separation of positive and normative claims aids in understanding arguments that others are advancing, in providing a better basis to decide whom to trust, to what degree, and on what questions, and in coming to an informed view of what decisions to favor. For policy debate overall, pursuing such separation of questions is likely to reduce confusion and conflict, and provide a sounder basis for seeking courses of action that might gain broad support.

To the extent that some distinction between positive and normative questions can be maintained, the two types of question are best dealt with in different ways. Positive questions – such as the evidence for present climate change, the changes that are likely over coming decades, and their consequences – are best examined by scientific processes, not democratic ones. For such questions, when a strong consensus exists among the relevant scientific experts, this is the closest thing we have to well-founded knowledge, and is entitled to substantial deference in policy debates. Strong expert consensus does not, of course, always exist on all policy-relevant positive questions. When it does not, the best indicator of the state of scientific knowledge is the range and distribution of judgment among relevant experts, to the extent this can be observed. When policy decisions have high stakes that depend on the answer to a positive question – e.g., how will climate change under different emission futures – policy actors can do no better than to take such a consensus or distribution of expert judgment as a true picture of present knowledge and uncertainty.

The problem with this advice, as discussed above, is that policy actors cannot reliably observe such a consensus or distribution of expert judgments themselves, but must rely on some type of scientific advisor or advisory process. Few policy actors have their own trusted scientific advisor, however, and in any case no individual advisor can provide authoritative advice on all questions from all fields of science. Consequently, there is a need for scientific advisory processes that multiple policy actors can rely on, to provide authoritative, trustworthy, and understandable answers to policy-relevant scientific questions. The process of synthesizing, evaluating, and communicating scientific knowledge to inform policy or decisions is called *scientific assessment*. There are many ways to organize and manage scientific assessment processes, but they typically involve assembling a committee of relevant experts under some competent and impartial organization. The committee is charged with reviewing current knowledge and uncertainty on specified policy-relevant questions, and producing a public report including summaries clearly understandable to non-specialists.

Scientific assessments connect the domains of science and democratic politics, but are distinct from both. They differ from science because, rather than advancing the active, contested margin of knowledge on questions that are important for their intrinsic intellectual interest, they seek to make consensus statements of present knowledge and uncertainty on questions that are important because of their implications for policy or decisions. They differ from democratic policy debate because they reflect deliberation over positive questions among scientific experts based on their specialized knowledge, not among all citizens or their representatives over what is to be done.

The need for effective scientific assessment to support environmental policy-making at both the national and international level has been widely recognized for at least 25 years. There are many ways to conduct scientific assessments, and many bodies that do them. The US federal government often calls on the National Academy of Sciences to advise it on scientific and technical matters related to national policy, and has sometimes established special scientific assessment bodies on particularly important or contentious issues. Scientific assessments can play an even stronger role in international environmental policy-making, because they can make authoritative statements of scientific knowledge that transcend differences in national policy positions. For example, atmospheric-science assessment panels on stratospheric ozone played a particularly influential role in reaching international agreements to control ozone-depleting chemicals, by making highly visible, authoritative statements on key scientific points.

Successful scientific assessments must skillfully navigate a path between the requirements of the domains of science and policy, satisfying criteria from both sides. They must assemble first-rank scientific expertise to provide an accurate and authoritative summary of current knowledge. They must be rigorously free of bias in their agenda, participation, processes, and outputs – and must absolutely avoid any hint of being constructed to favor a particular answer. Yet they must also communicate clearly, in terms that are simple enough to be understandable in policy debate, are relevant to decisions on the agenda, and are provided in time to be useful. Being useful to policy sometimes requires assessments to express judgments that would not be made explicitly within purely scientific processes, such as judgments of where the center of a scientific debate stands, the relative strength of various contending claims, or the confidence with which certain points are established. Such judgments are not normally expressed explicitly in scientific settings, because the scientific process can achieve such resolutions implicitly, and can wait for them, but providing a useful synthesis of present knowledge for policy may require making them explicit.

Assessments do not always succeed at making effective contributions to policy debates. They can fail to do so in many ways. For example, some assessments lose credibility by making explicit policy recommendations or otherwise going beyond their authoritative expertise. Others have failed to synthesize present knowledge into a coherent summary view, often out of reluctance to state explicit judgments of the relative strength of present contending claims or the distribution of present knowledge. There is no single model of how assessments can effectively manage the boundary between scientific and policy debate, but several areas of skill and judgment can make strong contributions to success. Leaders of successful assessments must maintain an alert ear to identify positive questions that policy actors perceive to be of high relevance. They must be able to motivate scientific participants to a level of synthesis and integration that is rarely done explicitly in purely scientific forums, while still maintaining rigorous standards of scientific debate. They must also exercise effective judgment to stay within their domain of expertise, and must be able to communicate clearly to a non-scientific policy audience without sacrificing scientific accuracy.

The Intergovernmental Panel on Climate Change, or IPCC, whose establishment we discussed in Chapter 1, is the primary body responsible for international scientific assessments of climate change. Since its establishment in 1989, the IPCC has undertaken four full-scale assessments of climate change – in 1990, 1995, 2001, and 2007 – as well as many smaller and more specialized reports. Each of the full assessments is a huge undertaking. The reports involve hundreds of scientists from dozens of countries as authors and peer reviewers, including many of the most respected figures in the field. These groups work over several years to produce each full assessment, and their reports are subjected to an exhaustive, publicly documented, multi-stage review process. In view of the number and eminence of the participating scientists and the rigor of their review process, the IPCC assessments are widely regarded as the authoritative statements of scientific knowledge on climate change. We will refer to these assessments repeatedly in summarizing present scientific knowledge throughout this book. In addition, in Chapter 5 we discuss in more detail how the IPCC's authoritative status was gained and how it can be defended and enhanced.

As we have emphasized, positive questions are not all there is to a policy debate. An effective scientific assessment process that pulls out relevant positive questions into expert forums still leaves various normative questions that must be addressed, explicitly or implicitly, to make decisions. These include, for example, how to evaluate different kinds of potential environmental changes with their associated risks and costs, including attitudes to uncertainty; how to trade off present against future harms; how optimistic to be about the potential for future technological change to ease the problem; and how to value the distributional tradeoffs associated with alternative policy choices. In contrast to positive questions about the climate, these questions are best dealt with through public deliberation and democratic decision processes. Indeed, the widespread pretense that current disagreements over climate-change policy basically arise from disagreements about the state of scientific knowledge has allowed policymakers to avoid dealing with their real responsibility, which is to engage these questions of political values in view of the present state of scientific knowledge, in order to decide what to do. The distinction between positive and normative questions cannot always be drawn cleanly in practice, but we contend that it is possible to disentangle these much more than has been done in the present debate, and that efforts to do so are likely to yield a more informed and less contentious policy debate, and perhaps even assist in the identification of widely acceptable policy choices.

The remaining chapters follow our ambition to distinguish positive from normative questions. Building on the brief scientific primer in Chapter 1, Chapter 3 identifies the most important positive questions about the climate for the present policy debate and summarizes the present state of scientific knowledge about them and includes a discussion of several widely circulated contrary claims. Chapter 4 examines alternative policy responses to the climate-change issue, concentrating on present knowledge about what options are available, how effective they are likely to be, and what their costs and other consequences are likely to be. Chapter 5 outlines the current state of climate-change politics, then returns to our central concern about partisan distortion and misuse of scientific knowledge and uncertainty in the policy debate. We outline a few prominent instances of such argument, and discuss in more detail how effective scientific assessment processes could reduce the latitude and the incentives for such misrepresentation. Finally in Chapter 5, we step partway into the role of advocates ourselves, and provide a broad outline of an approach to climate-change policy that holds the hope of breaking the present deadlock and securing wide agreement.

Further reading for Chapter 2

Alexander Farrell and Jill Jäger, eds. (2005). *Assessments of Regional and Global Environmental Risks: Designing Processes for the Effective Use of Science in Decision-making*. Washington, DC: Resources for the Future.

> The studies collected in this volume examine how scientific and technical assessments of several environmental issues were designed and managed, to provide practical insights into how to conduct effective assessments. Produced from the same research project as the Mitchell et al. volume cited below, this one uses the same framework to evaluate assessments according to their scientific credibility, legitimacy, and decision relevance and saliency.

Sheila Jasanoff (1990). *The Fifth Branch: Science Advisors as Policymakers*. Cambridge: Harvard University Press.

> This study of scientific advisory bodies to US government agencies examines the processes by which the boundaries between the scientific and political domains were negotiated in several regulatory controversies, and the conditions that contributed to more or less stable and effective maintenance of the boundary – and a more or less constructive relationship between scientific advice and regulatory decision-making – in each case.

Thomas Kuhn (1962). *The Structure of Scientific Revolutions*. Chicago: University of Chicago Press.

> This study of the social processes by which scientific disciplines make progress was the first to note the contrast between two starkly different modes of change in scientific understanding: normal, incremental progress that depends on a certain deep and unexamined structure of shared assumptions (which Kuhn called "paradigms") about what questions are important and what lines of research are interesting or promising; and occasional revolutionary upheavals that follow the accumulation of some critical quantity of results that do not fit within the paradigm.

Ronald Mitchell, William Clark, David Cash, and Nancy Dickson, eds. (2006). *Global Environmental Assessments: Information and Influence*. Cambridge, MA: MIT Press.

This synthesized collection of studies of scientific assessment processes in several environmental issues examines the mechanisms by which assessments can inform and influence policy decisions, and the conditions that shape whether they do so. It examines how assessments operate and are used in political settings, and argues that effective assessments, in addition to being scientifically credible, must also be perceived as legitimate in their process and participation, and must present their outputs in terms that are sufficiently salient and relevant to decision-makers.

William D. Ruckelshaus (1985). Risk, Science, and Democracy, *Issues in Science and Technology*. 1:3, Spring 1985, pp. 19–38.

Ruckelshaus, Administrator of the US Environmental Protection Agency under Presidents Nixon and Reagan, argues that effective environmental policy depends on maintaining enough separation between scientific-based processes of assessing environmental risks, and more political processes of deciding what to do about the risks based on the best scientific information that is available.

3

Human-induced climate change: present scientific knowledge and uncertainties

The scientific primer in Section 1.2 discussed the basic physics of climate and why the climate is expected to warm if more greenhouse gases are added to the atmosphere. This chapter deepens the discussion, examining current scientific knowledge of the observed changes in the Earth's climate, the extent of human influence on these changes, and potential future changes in the climate. Contrary to the impression you might get from following the debate in the news, current scientific understanding of the climate, its variations, and influences on it, is actually quite advanced. We parse the questions of the reality and importance of climate change into four separate, specific questions.

* *Is the climate changing?*
* *Are human activities responsible?*
* *What further climate changes are likely?*
* *What will the impacts be?*

Sections 3.1 through 3.4 review the available evidence and summarize present scientific knowledge on each of these questions, as well as key uncertainties and controversies. Section 3.5 reviews a few of the most widely circulated claims that deny the main points of current scientific knowledge about climate change. Mostly circulated in non-scientific forums, these claims are usually presented by those who oppose policy action to limit climate change, sometimes called "climate change skeptics."

3.1 Is the climate changing?

To determine whether the climate is changing, we must be precise about what we mean by "climate." Recall from Chapter 1 that climate describes average meteorological conditions for some place and time period of interest – e.g., average November conditions for Washington, DC between 1970 and 1990 – as opposed to the weather of any particular day. Climate includes temperature, and also other factors such as humidity, precipitation, cloudiness, winds, etc. And it includes not just average values of these factors, but the statistics of variations from the average. Changes in variation or extremes may in fact matter more than changes in averages. For example, we may well care more about the occurrence of extreme heat waves than about changes in the average summer temperature.

Although changes in any climatic characteristic can have important impacts, we focus on temperature because it is the aspect of climate with the best data, the one with the strongest theoretical connection to greenhouse-gas emissions, and the one that principally drives other changes. There is much evidence that other aspects of climate, such as precipitation patterns, are also changing, but we will not discuss it here. Our focus will be on the past few decades to few centuries, because this is when human activities have been increasing atmospheric greenhouse gases. We will also briefly discuss climate changes over the past 100 million years – not because humans had any effect on this, but to put recent changes in context.

Finally, the place where we will look for changes is in the average surface temperature of the entire Earth, averaged over the entire year. We do this not because this global average matters to anyone – no one lives in the global average, and people care far more about specific changes where they live – but for practical reasons. The global average is where temperature trends are easiest to identify. Regional climate events such as severe droughts or heat waves, or strong short-term trends, can occur by chance even in a climate that is not changing, so unless the overall frequency or intensity of such events is increasing they do not necessarily provide evidence of global climate change. Looking at the global average is more reliable because these smaller-scale regional variations tend to average out.

To determine if the Earth's surface is warming, we need measurements of temperature or some related quantity over a long enough period to establish a trend. There are many different sources of relevant data to draw on, each with distinct strengths and weaknesses. We will review several of the most important of these data sources, and show that they paint a consistent picture of rising temperatures. Considered together, these sources provide decisive evidence that the Earth's surface has been warming for the last few hundred years, with particularly rapid warming over the last few decades of the twentieth century.

3.1.1 The surface thermometer record

The simplest way to measure the temperature of the Earth is to place thermometers – such as simple liquid-in-glass thermometers like the one you

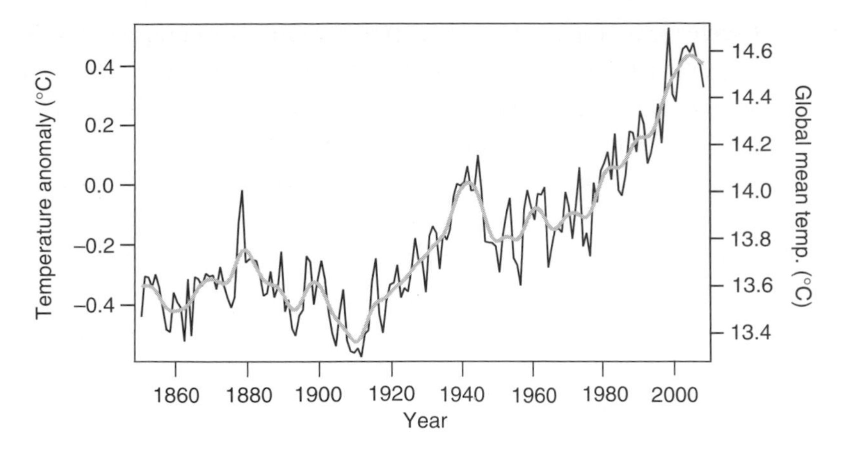

Figure 3.1 Global and annual average surface temperature anomalies (°C) from 1850 to 2007, measured relative to the 1961–1990 average. The black line is the annual average, while the gray line is smoothed to show longer-term variations. *Source*: data from the HadCRUT3v data set, Hadley Centre for Climate Prediction and Research.

may have on your back porch – in many locations around the world, and record the temperature at each location every day. By combining measurements taken all over the globe, you can construct an estimate of the average surface temperature of the Earth. People have been making these measurements at thousands of locations, both on land and from ships at sea, for about 150 years, and these temperatures are plotted in Figure 3.1. The record shows that from 1906 to 2005, the average surface temperature of the Earth rose by 0.74°C. Most of this increase occurred in two distinct periods, from 1910 to 1945 and from 1976 to the present, with a small cooling between these periods and many shorter-term bumps and wiggles over the century. (Section 3.2 will discuss the origin of the cooling period and the bumps and wiggles.) The last two decades have been the warmest since measurements began in the mid-nineteenth century, and the last eleven years (1997 through 2007) are the eleven warmest individual years in the record.

Figure 3.1 plots temperature *anomalies* rather than actual temperatures. The temperature anomaly is the difference between the actual temperature each year and some reference temperature. In this figure, the reference is the average temperature between 1961 and 1990, about 14°C. The figure shows that temperatures between 1860 and 1920 were 0.2–0.4°C below this 1960–1990 average, while temperatures over the past 10–20 years have been 0.5°C above this average.

Why show anomalies rather than actual temperature measurements? The main reason is that many sources of temperature-related data, such as the glacier data described in the next section, can only show changes over time, equivalent to temperature anomalies, not absolute temperatures. Even if a particular data set could be expressed as absolute temperatures, global temperature data are usually expressed as anomalies so records from all sources can be compared. In addition, absolute temperature can vary over short distances, such as between a city and a nearby rural area, or between two nearby sites at different altitudes. Anomalies, however, tend to be constant over longer distances, making the calculation of anomalies easier and requiring a less dense measurement network.

The surface thermometer record provides the strongest evidence that the Earth is warming, and the most accurate estimate of how much it has warmed. Why is this data set so good? The main reason is that these are the most direct measurements of the Earth's temperature. Other methods of determining temperature trends are *indirect*. They do not measure surface temperature itself, but infer it from some other quantity such as the length of glaciers or the extent of sea ice. For these indirect data sets, converting changes in the observed quantity to a temperature trend introduces additional uncertainty. In addition, the technology behind the thermometer is hundreds of years old. Such technical maturity adds great confidence that the trend observed in the temperature is real and not some undiscovered artifact of the measurement instrument.

Despite its strengths, this data set still has limitations. The 150-year history of continuous observations is in some respects an advantage, but changes in how observations were made over this long period can also introduce errors. To illustrate how these can occur, consider a hypothetical temperature station that has operated from 1861 to the present. In 1861 it was operated by a farmer, who read a liquid-in-glass thermometer and recorded the temperature each day at noon. While thermometer technology was mature even then, there were occasional errors in the record, because of instrument problems (for example, a bubble in the thermometer), or because the farmer misread or mis-recorded the day's temperature. Simple errors like these are unimportant for estimating long-term trends, because they are as likely to go in one direction as the other and so average out in the long term.

When the farmer died in 1890 his son continued the daily temperature readings, but he made them at 3:00 PM instead of noon. Since it is usually warmer in mid-afternoon, temperatures recorded at this station suddenly increased. In 1902, the barn next to the thermometer burned down and the thermometer was moved to a south-facing hillside that received more sunlight, so the recorded temperatures increased once again. Over the next 50 years, the nearby city grew until it eventually surrounded the farm. Cities are warmer than the surrounding countryside, because roads and buildings are darker than vegetation and so absorb more sunlight – a phenomenon known as the "urban heat island effect" – so this urban sprawl caused an additional warming trend in the record. These errors, unlike simple reading and recording errors, can introduce spurious trends in the temperature record.

These types of error are well known, and various techniques are used to identify and correct them. Changes in observing practices such as changing the measurement time from noon to 3:00 p.m. or relocation of the thermometer can be identified by looking for sudden jumps in a station's record, then checking its log books to see what changed on that date. Once the cause is identified, the station's prior records can be adjusted to account for the change in observing practices. The size of the urban heat island effect can be estimated by comparing a station in a growing urban area with a nearby rural station. While the urban heat island effect can be important in estimating local or regional trends, it is not a major factor in the global trend shown in Figure 3.1. Urban areas make up a small fraction of the Earth's surface area, so the trend calculated using rural stations alone is very similar to that calculated using all observing stations.

The final problem with the surface thermometer temperature record concerns how thoroughly and uniformly the observing stations cover the Earth's surface. The coverage is extensive, but far from complete. Most stations are located where people live or travel, so most measurements are made on land, in densely populated regions. Coverage is thin in polar regions, uninhabited deserts, and ocean regions outside major shipping lanes. In addition, coverage has changed over time, especially on the ocean. If newly added regions are on average warmer or cooler than regions previously observed, this could also create a spurious trend. As with changes in observing practices, scientists are aware of these problems and have developed techniques to determine a robust average temperature from sparse and changing coverage, and to estimate how much bias might still remain in the record. For example, comprehensive measurements of sea surface temperature by satellite can show what errors were caused by the earlier sparse coverage of ocean measurements and the expansion of ocean coverage over time.

The surface thermometer record is the most important historical data set used in studies of climate change. The data set has been extensively studied, the potential errors in it are well known and well understood, and adjustments have been made where needed to correct these errors. Some uncertainty remains, and this is reflected in an uncertainty estimate of plus-or-minus 0.18°C in the 100-year warming trend of 0.74°C between 1906 and 2005.

But for all the strengths of this record, the conclusion that the Earth is warming does not rest on it alone. As Chapter 2 stressed, scientific claims of such importance must be verified, ideally in several independent ways, before being widely accepted. The rest of this section discusses other data that provide independent estimates of how the Earth is warming.

3.1.2 The glacier record

In cold regions, such as near the poles or at high elevations in mountains, snow that falls in winter does not all melt the next summer. Under the right conditions, snow can accumulate to great thickness over many years, compressing under its own weight to form a thickened sheet of ice known as a glacier. Glaciers cover about 10 percent of the Earth's land area, mostly in Antarctica and Greenland, and contain the vast majority of the world's fresh water.

Changes in the size of a glacier provide a good measure of local temperature changes. Local warming will generally increase melting and cause a glacier to shrink or retreat, while local cooling will generally cause it to grow or advance. Lengths of many glaciers have been measured for hundreds of years, so looking at glaciers all over the Earth can provide data on changes in global temperature.

Historical records show a clear pattern of retreating glaciers. Of the 36 glaciers that were monitored between 1860 and 1900, only one advanced and 35 retreated. Of the 144 monitored between 1900 and 1980, two advanced and 142 retreated. Figure 3.2 shows changes in average glacier length since 1700 for five world regions. It shows that glaciers began retreating around 1800, with the shrinkage accelerating later in the nineteenth century. The pattern of glacier retreat over the past century or so is consistent worldwide, with all five regions showing similar trends.

Warming is the most obvious explanation for this worldwide retreat, but other possible explanations must also be considered. The balance between snow accumulation and melting, which fundamentally determines a glacier's expansion or retreat, could also be altered by a decrease in snowfall, or by a decrease in cloudiness that allowed more sunlight to fall on the glacier. Models of glacier dynamics, called "mass balance models," can calculate the expected changes in glacier length from each of these factors.

For a typical mid-latitude glacier, these models find that it would require a 30 percent decrease in cloudiness or a 25 percent decrease in annual snowfall to make the same retreat as a 1°C warming. Such large changes in snowfall or cloud cover could occur locally or regionally, but worldwide trends this large over a century are highly unlikely and have not been observed. For this reason, glaciologists consider a warming trend to be the dominant cause of the observed glacier retreat. Moreover, the warming calculated from the observed retreat is consistent with that observed in the surface thermometer record, providing independent verification of the warming trend.

One limitation of glacier data is coverage. Glaciers only occur in cold places, so a temperature trend calculated from glacier retreat – while consistent with the global thermometer trend – tells only part of the story. We will see below,

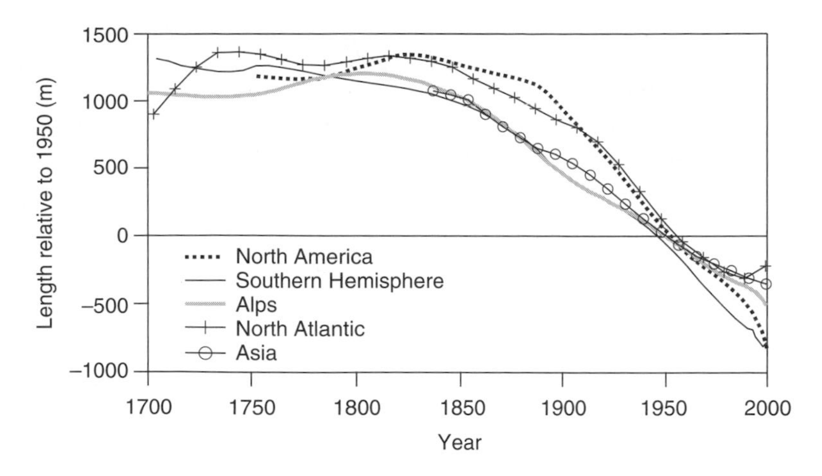

Figure 3.2 Change in mean glacier length over time, measured relative to length in 1950, for five world regions.
Source: after Fig. 4.13, of IPCC [2007a].

however, that other data with different regional coverage provide similar confirmation of warming trends, and so strengthen confidence that the observed warming is truly global.

3.1.3 Sea level

When the climate warms, sea level rises for two reasons. First, like most substances, water expands when it warms, so warming increases the volume of the water in the oceans. Second, when warming melts glaciers or other ice on land, the melt water runs into the oceans and further raises their level. (Melting sea ice does not change sea level, because the ice is floating.) The opposite effect occurs during ice ages. At the peak of the last ice age, the immense volume of water stored in continental ice sheets lowered sea level 120 m below the present level.

Non-climate processes can also affect sea level, complicating attempts to infer a temperature trend from sea level data. For example, sinking of coastal land can make local sea level appear to rise, even if the absolute sea level is constant. Such sinking can arise from slow natural movements of the Earth's crust, or from human activities such as groundwater extraction. We have good knowledge of where such sinking is happening, however, and adjust for this in interpreting local sea-level measurements to calculate a global average, as in Figure 3.3.

Figure 3.3 shows trends in global-average sea level since the late nineteenth century, calculated from a combination of tide gauge and satellite measurements. Over the twentieth century, the rate of increase was about 1.5 mm per

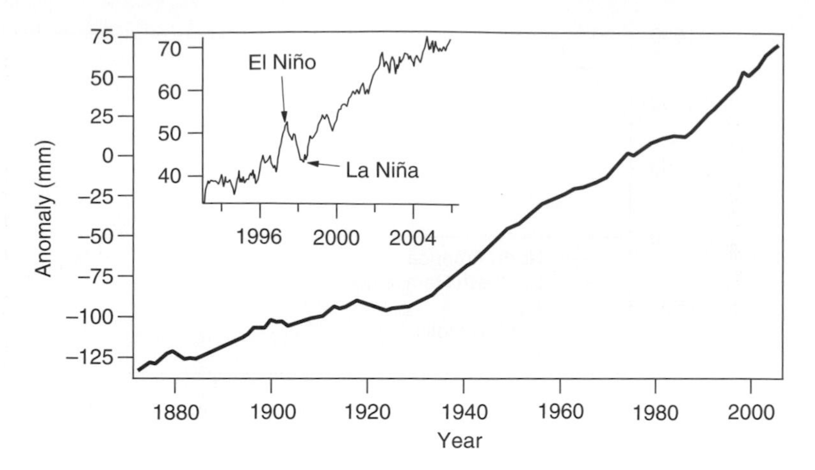

Figure 3.3 Global and annual average sea level anomaly, measured relative to the 1961–1990 average and smoothed to show decadal variations. The inset shows a close-up of unsmoothed 1993 to 2006 anomalies.
Source: adapted from Figures SPM.3 and 5.14 of IPCC [2007a].

year, or 15 cm total over the century. The few records that extend back into the nineteenth century suggest that sea level rose faster in the twentieth century than in the nineteenth. Over recent decades, the rate of rise has increased. In the last 40 years, the increase has been about 1.8 mm per year, of which thermal expansion accounts for about one-quarter (0.42 mm/yr). In the most recent ten years of data, from 1993 to 2003, the increase has been 3.1 mm per year, with thermal expansion accounting for about half (1.6 mm/yr) and melting of land-ice the remainder. Studies of the individual processes driving sea level rise show that the observed changes in sea level are consistent with the observed rate of warming found in other data sets.

As with all data sets, there are uncertainties in this one. For example, there is a possibility that changes in the amount of water stored on land in forms other than ice, such as lakes and aquifers, might also be contributing to sea level changes. Although quantitative estimates of the contribution of such sources are quite uncertain, they probably do not make an important contribution to the observed sea level trend.

3.1.4 Sea ice

At the cold temperatures found in polar regions, seawater freezes to form a layer of ice on the top of the ocean, typically a few meters thick. The area covered by sea ice varies over the year, reaching a maximum in late winter and a minimum in late summer.

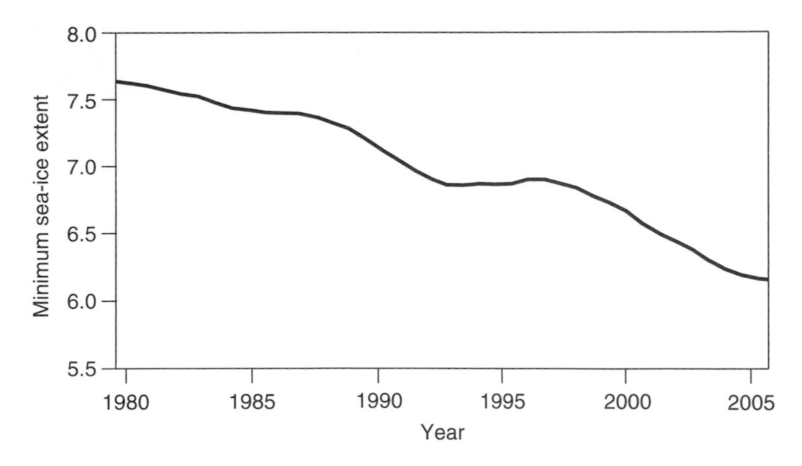

Figure 3.4 Satellite measurements of summer minimum Arctic sea-ice extent (millions of sq. km), smoothed to reveal decadal variations.
Source: Fig. 4.9, IPCC [2007a].

Given the rapid warming now occurring, we might expect to see changes in sea ice – and we do. Figure 3.4 plots Arctic sea-ice area, measured at the late-summer minimum, over the past three decades. There is a clear downward trend, with a reduction in ice-covered area of 1.5 million square kilometers between the late 1970s and 2005, an average reduction of 7.4 percent per decade. In summer 2007, so much ice melted that an ice-free passage opened between the Atlantic and Pacific through the Arctic islands of northern Canada. Reductions in sea-ice area averaged over the whole year, as opposed to the summer minimum, show a smaller rate of loss of 2.7 percent per decade.

In addition to shrinking in area, sea ice has also grown thinner. Measurements from submarines show that from 1987 to 1997, sea-ice thickness in the central Arctic decreased by up to 1 m – a loss of about one-third of the average thickness. Together, these sea-ice measurements provide strong confirmation of the warming in the Arctic region measured by the surface thermometer network.

The Antarctic is a different story. Sea-ice area around that continent has remained stable since the mid-1970s. It is unclear whether this is consistent or not with Antarctic temperatures – sparse and uncertain temperature data provide no clear picture about whether Antarctica has warmed or not over the past few decades.

3.1.5 Ocean heat content

One of the most confident predictions of global-warming theory is that most of the energy trapped in the atmosphere by greenhouse gases will end

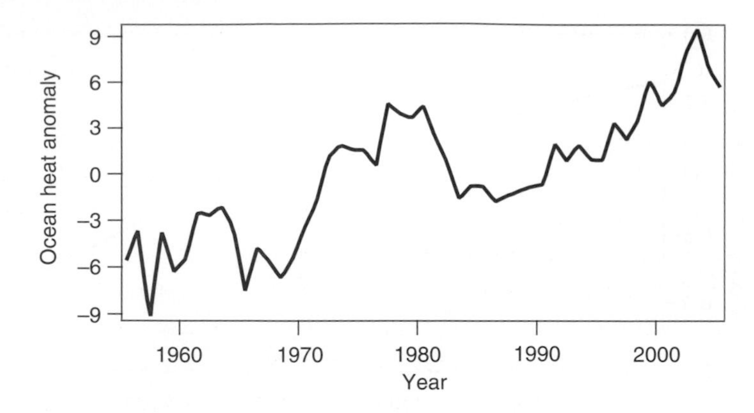

Figure 3.5 Heat content anomaly (in units of 10^{22} J) for the top 700 m of the ocean, measured relative to the 1961–1990 average.
Source: adapted from Fig. 5.1, IPCC [2007a].

up heating the ocean. To test this prediction, millions of ocean temperature profile measurements have been made in order to calculate changes in the heat content of the oceans. Results of this calculation are shown in Figure 3.5. While there are strong variations over years and decades, it is clear that the top layer of the oceans is experiencing a long-term heating trend.

3.1.6 Satellite temperature measurements

Since 1979, weather satellites have provided a new, independent source of data about global temperatures. One particular satellite instrument, the Microwave Sounding Unit (MSU), measures microwave radiation emitted by the atmosphere, from which the average temperature at various altitudes in the atmosphere can be calculated. As a measure of global-average temperature, satellite data have the great advantage of covering the entire Earth, including oceans and uninhabited areas, so there is no risk of partial or biased coverage.

The data also have some serious limitations, however. The record covers only about three decades, a short period from which to estimate trends. Moreover, the record is stitched together from data gathered by about a dozen satellites, each of which lasts just a few years before it fails and is replaced by the next. Because only one instrument is working at any given time, problems with that one instrument can lead to significant errors in the inferred trends. This stands in sharp contrast to the surface thermometer record in which measurements are made on thousands of thermometers each day. Problems with any single thermometer will have a negligible effect on the long-term trend.

It also makes the estimated temperature trend highly sensitive to how the records from successive satellites are intercalibrated. To understand this, suppose you are trying to watch your weight, but your scale breaks and a month passes before you buy a new one. If the new scale says you are 2 pounds heavier than your last reading on the old one, does this mean you have gained 2 pounds? Or does the new scale just read 2 pounds heavier than the old one? You could tell which of these was the case if you bought a new scale before the old one broke, and measured yourself on both scales for a while to estimate the difference between them – if you had the foresight, patience, and money to do this. The weather service, which operates these satellites, has tried to do this, by launching each new satellite while the previous one is still operating and so providing overlapping measurements for long enough to calibrate the new instrument. But since it is impossible to predict exactly when an instrument is going to fail, they have not been entirely successful in obtaining long enough overlapping records. As a result, the temperature trend estimated from the satellite data is quite sensitive to how you connect data between these satellites.

This intercalibration problem is just one of several difficult problems with the satellite record. Subtle changes in the satellites' orbits and drift of the calibration of the satellites' sensors can introduce errors much larger than the actual trend. These problems are not an issue for the surface thermometer record, which uses centuries-old technology to make its measurement.

A final concern with the satellite temperature record is that it does not measure the temperature at the Earth's surface, but the average temperature of a fairly thick layer of the atmosphere. The most widely quoted satellite measurement is the average temperature of the lower troposphere, extending from the surface to about 8 km altitude (roughly where commercial jets fly). Trends in this atmospheric layer are not necessarily directly comparable to trends in the surface record.

Figure 3.6 shows global-average temperature anomalies of the lower troposphere calculated from the satellite record by two independent scientific groups. Also shown is the temperature anomaly for the same range of altitudes, measured from weather balloons. The three data sets all show a similar warming trend as well as similar year-to-year variations, which are driven mainly by internal variations like El Niño (see box below). For the period from 1979 to 2004, the UAH group calculates a warming trend of 0.12 ± 0.08°C per decade, while the RSS group calculates 0.19 ± 0.08°C per decade. These two groups are using the same raw data from the MSU satellite instruments, so the difference in their calculated trends comes entirely from different assumptions about technical details of the calculation, including calibration between satellites. Over the same period (1979 to 2004), the balloon measurements show a trend of

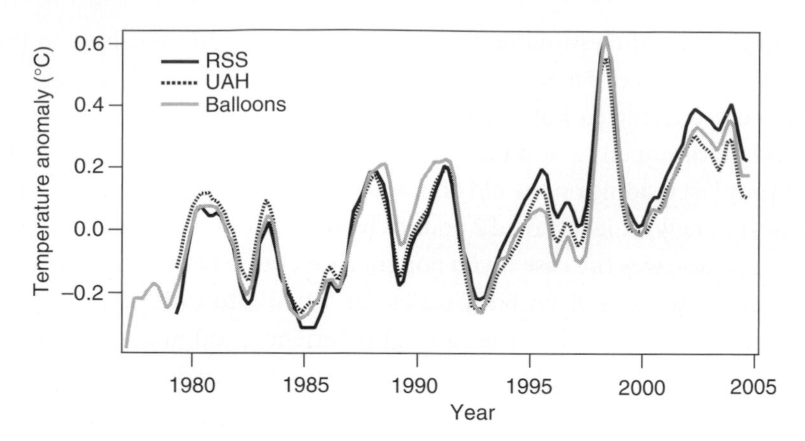

Figure 3.6 Time series of globally averaged lower atmospheric temperature anomalies, measured relative to the 1979–1997 average. RSS is Remote Sensing Systems, UAH is University of Alabama at Huntsville. The gray line shows balloon-borne thermometer measurements from the Hadley Centre (HadAT2 data). *Source*: Fig. 3.3a of CCSP [2006].

0.14 ± 0.07°C per decade, while over the full period that balloon measurements are available (1958 to 2004), their trend is 0.16 ± 0.04°C per decade.

These three records of temperature trends in the lower atmosphere all show warming over the past three decades, and agree with each other to within their stated uncertainties. There were formerly larger discrepancies between the satellite and surface records, which were sometimes used to deny the reality of the observed surface-thermometer warming trend. Repeated improvements in the records, however, have made the remaining discrepancies in global temperature trends relatively small. Discrepancies still exist in certain altitude and latitude regions, and interpreting and reducing these data is an active area of current scientific research. But at global scale, these atmospheric measurements are now broadly consistent both with each other, and with the surface thermometer trend of 0.1–0.2°C per decade over the same period.

What is El Niño?

Some of the year-to-year changes in the global-average temperature record, such as the large wiggles in Figure 3.6 and in the unsmoothed line in Figure 3.1, come from known patterns of climate variability that occur over periods of a few years to a few decades. Of these, the most important is El Niño, one phase of a large-scale atmospheric oscillation more generally called the *Southern Oscillation*. The Southern Oscillation is a sloshing of warm surface water back and forth across the tropical Pacific Ocean every few years. The Southern Oscillation has two phases, called El Niño and La Niña, each of which lasts a year or two. In addition, the oscillation sometimes lingers in a neutral phase between these states.

In the neutral phase, equatorial trade winds blowing to the west push warm tropical water into the Western Pacific, causing upwelling of cool water off the west coast of South America. During an El Niño, the trade winds slacken, and the warm water moves eastward to fill the entire tropical Pacific basin. Linked changes in temperature and rainfall extend worldwide, and the Earth's average temperature increases. The size of this effect is illustrated by the El Niño of 1998, the strongest of the twentieth century. Figures 3.1 and 3.6 show how global-average temperatures that year were several tenths of a degree warmer than the next and prior years. Warm temperatures that year also caused the ocean to expand, producing the surge in sea level that can be seen in the inset of Figure 3.3.

In the La Niña phase, the trade winds strengthen, pushing cool water westward to fill much of the tropical Pacific. This also causes worldwide changes in temperature and rainfall opposite to those seen during an El Niño, including a cooling of the Earth's average temperature and a drop in sea level – as can be seen in the same figures above during the La Niña of 1999.

3.1.7 Climate proxies

All the records discussed thus far involve measurements of temperature, or something associated with temperature, in real time. Each data source provides some information about the climate, but only for as far back in time as it has been measured. At best, these only provide information back a few hundred years, and most cover a considerably more recent period.

Another type of information about past climates is provided by *climate proxy* records. A climate proxy is a record of past climate that has been imprinted on some long-lived physical, chemical, or biological system. Different proxies, depending on how long they last, can extend climate records back hundreds, thousands, even millions of years into the past. This section discusses several of the most important and widely used climate proxies, and what they tell us about past climate variation.

Tree rings

Tree growth follows an annual cycle, which is imprinted in the rings in their trunks. As trees grow rapidly in the spring, they produce light-colored wood; as their growth slows in the fall, they produce dark wood. Figure 3.7

Figure 3.7 Cross-section of a tree, showing its rings. Note how the rings vary in size and density. Photo courtesy of the Laboratory of Tree-Ring Research, University of Arizona.

shows a cross-section of a tree trunk and its rings. Because trees grow more, and produce wider rings, in warm years, the width of each ring gives information about climate conditions around the tree in that year. The rings of a long-lived tree can provide temperature data extending back hundreds of years.

The key to using tree rings as a climate proxy is finding a quantitative relation between ring width and temperature in the tree's location. This is done by examining rings from recent years for which thermometer records are also available. A relationship between ring width and temperature is estimated for this period, which can then be used to estimate temperature for that part of the tree's life before temperature was measured directly.

There are a few difficulties to using tree rings as a climate proxy. First, it is hard to separate temperature's effect on tree growth from that of other climate variables like rainfall. This is generally handled by carefully selecting trees in regions where, for the past century or so, tree growth has been primarily responsive to temperature and insensitive to precipitation.

Second, the reconstructed temperature records assume the relation between ring width and temperature estimated from the recent past applies over the entire life of the tree. Finally, like many of the data sources we consider, tree-ring reconstructions are only available for a fraction of the Earth's surface. They are obviously not available over oceans, or from desert or mountainous areas where no trees grow. They are also not available in the tropics, where the weaker seasonal cycle causes trees to grow year-round, so they do not produce rings.

Ice cores

We discussed above how the advance or retreat of mountain glaciers gives information about temperature changes over the past few centuries. In addition, the chemical and physical characteristics of glacial ice provide a rich store of information about conditions at the time the snow fell. The chemical composition of the ice, such as the fraction of heavy isotopes[1] of hydrogen and oxygen, can be used to infer the air temperature around the glacier when the snow fell, as can variations in the size and orientation of ice crystals. Small air bubbles trapped when glacial ice formed preserve a snapshot of the chemical composition of the atmosphere at that moment. In addition, the dust trapped in the ice gives information about prevailing wind speed and direction, and about how wet or dry the regional climate was when the ice formed, because more dust blows around during droughts. Finally, because sulfur is one of the main effluents of volcanoes, measurements of sulfuric acid in glacial ice can show whether there was a major volcanic eruption around the time the ice formed.

Researchers retrieve a time series of all this information by drilling down into the ice sheet with a hollow drill bit and removing a long column of ice, a few inches in diameter, known as an ice core. The deeper the ice, the longer ago the ice was deposited and the further back the time for which it provides climate information.

Reconstructing past climate information from an ice core requires two steps. First, the age of each ice layer must be determined from its depth inside the glacier. Much effort has been spent on this problem, because the rate of ice accumulation varies over time and because ice inside the glacier can compress and flow under the great weight of the ice above. Second, the characteristics actually observed, such as the abundance of heavy isotopes of hydrogen, must be translated into the climatic characteristics of interest, in this case, temperature. Ice cores from the thickest, oldest ice sheets in Antarctica and Greenland have provided climate reconstructions dating back an amazing 800,000 years.

Corals

Corals are small marine animals that live in colonies anchored to reefs in warm ocean waters, mostly in the tropics. The reefs, which are made up of skeletons of previous generations of coral, can be thousands of years old. Much

[1] Isotopes are different forms of atoms that differ in the number of neutrons in their nucleus. Molecules containing different isotopes have slightly different chemical and physical properties.

like tree rings, coral skeletons grow outward in bands that can be dated. The chemical composition of the reef can provide information about past climate and ocean conditions, including ocean temperature, precipitation, salinity, sea level, storm incidence, and volume of nearby freshwater runoff. Obtaining past climate information from corals requires a process similar to that used for ice cores. First, a drill is used to obtain a column of coral material. The precise date of each layer of coral is then determined. Finally, the chemical composition of the coral is measured to provide a snapshot of conditions around the coral at that time.

Ocean sediments

Billions of tons of sediment accumulate at the bottom of the ocean every year. Like ice cores and corals, this sediment contains information about climate conditions at the time it was deposited. The most important source of information in sediments comes from the skeletons of tiny marine organisms. The relative abundance of species that thrive in warmer versus colder waters gives information about surface water temperature. The chemical composition of the skeletons and variations in the size and shape of particular species provide additional clues. In total, ocean sediments can provide information about water temperature, salinity, dissolved oxygen, nearby continental precipitation, the strength and direction of the prevailing winds, and nutrient availability, back for as long as a hundred million years.

Boreholes

Temperatures measured today at different depths underground provide a different way to infer how surface temperature varied in the past. To understand how this works, think about cooking a frozen turkey. You can tell how long the turkey has been in the oven by measuring the temperature at different depths below its skin. If the turkey is hot on the surface but still frozen just below the skin, then it has been cooking for only a short time. If the center of the turkey is 165°F, then it has been in the oven for several hours – and you should take it out before it is overcooked! In an analogous way, measuring the temperature of the Earth at many depths in deep narrow holes called boreholes allows you to infer the history of the ground surface temperature over the past few hundred years.

One difficulty in interpreting borehole temperatures is that they provide information about the temperature of the ground, while surface thermometers report the temperature of the air a few meters above the ground. The difference in temperatures derived from these two sources is usually small, but can be significant depending on properties of the surface such as land-cover, soil moisture, and winter snow cover. In central England, for example, where the ground is rarely snow-covered and major land-use changes have not occurred for several centuries, surface temperature trends inferred from boreholes are very similar to those in thermometer records. But in northwestern North America, borehole estimates of surface warming over the twentieth century are 1–2°C larger than the warming in the thermometer record, probably due to changes in land-use and average snow-cover over the century. Such discrepancies are considered and controlled to the extent possible in order to construct a consistent historical temperature record.

Past climate from proxy data: an integrated picture

Each climate proxy provides a different view of climate history. For example, tree rings provide information about mid-latitude climate on land dating back a few hundred years; corals provide information about the tropical sea temperatures dating back a few thousand years; and ice cores provide information about climate in polar regions dating back a few hundred thousand years. While each data source has its unique limitations and uncertainties, and its own limits in time and space, putting them all together provides a picture of the Earth's climate dating back 100 million years, with some records extending even further.

We will examine what these proxy records show about Earth's past climate variation in four views, starting with the longest view and then moving progressively closer to the present time. Figure 3.8 shows a reconstruction of polar temperatures over the last 70 million years, derived from ocean sediments. The warmest temperatures in this record occurred about 50 million years ago, or 15 million years after the extinction of the dinosaurs, in a period called the Eocene climatic optimum. During that time, the planet was far warmer than today. Forests covered the Earth from pole to pole, and plants that cannot tolerate even occasional freezing lived in the Arctic, along with animals such as alligators that today live only in the tropics. Since that time, the Earth has experienced a significant long-term cooling.

Figure 3.9 zooms in closer, to show global temperature variation over the past four million years. Like the 70-million year record in Figure 3.8, this record also shows a general cooling trend. Starting around three million years ago, about when large ice sheets like that on Greenland first appeared in the Northern Hemisphere, it also shows the appearance of large oscillations between warmer and cooler periods. In the cool periods, called *ice ages*, the ice sheets expanded to cover large parts of the Northern Hemisphere. In the warm periods between the ice ages, called *interglacial periods*, the ice sheets contracted. From about two and a half million to one million years ago, ice ages occurred every 41,000 years.

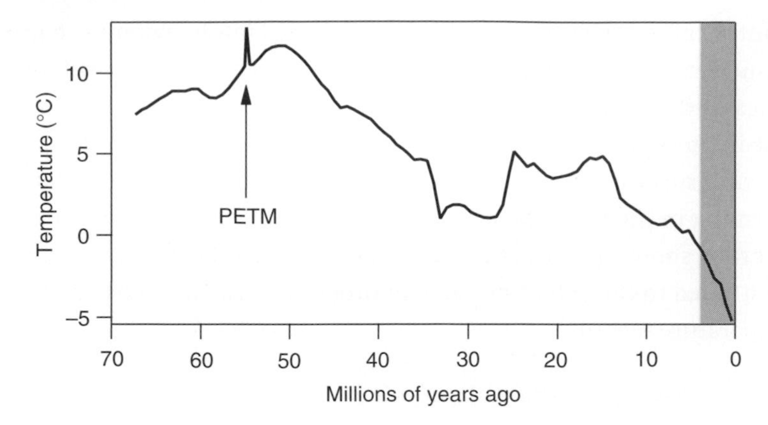

Figure 3.8 Reconstructed temperature of the polar regions over the last 70 million years. The sharp temperature spike 55 million years ago is called the Paleocene-Eocene thermal maximum or PETM. The gray bar on the right shows the last 4 million years, which are expanded in Fig. 3.9. Because this time series is from ocean isotopes, it is sensitive both to temperature and to the total volume of ice on land. Starting about 35 million years ago, some of the variation here comes from land ice volume rather than temperature. The overall trend, however, mostly represents changes in temperature.
Source: after Fig. 2 of Zachos et al. [2001].

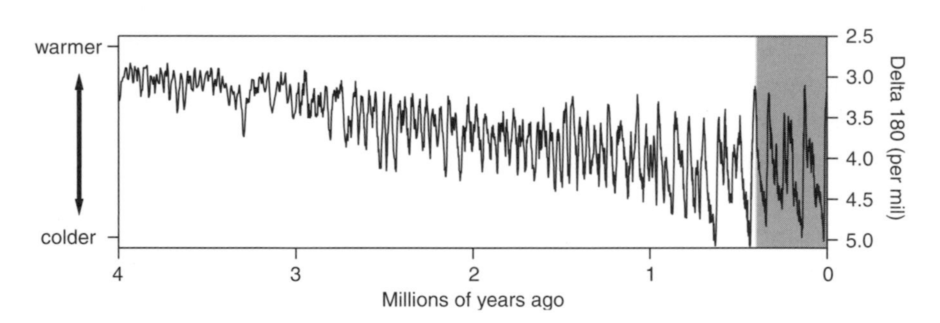

Figure 3.9 Measurement of global-average relative temperature over the last four million years. The vertical axis measures relative abundance of the heavy isotope of oxygen-18, a proxy for temperature, in ocean sediment cores. The temperature difference between the top and bottom of the graph is ~10°C. The gray bar on the right shows the last 410,000 years, which are expanded in Fig. 3.10.
Source: based on the analysis of Lisiecki and Raymo [2005].

Since then, for reasons that are not well understood, the frequency of ice ages reduced to once every 100,000 years.

Figure 3.10 zooms in again, showing a record of temperature and CO_2 for the Antarctic region over the last 410,000 years constructed from ice cores. This

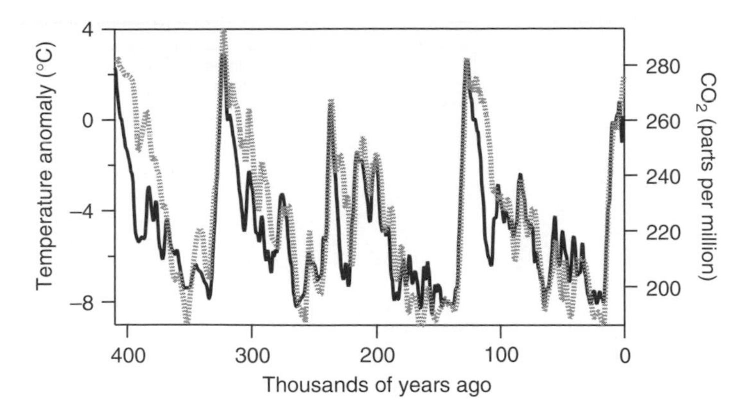

Figure 3.10 Temperature anomaly of the southern polar region (black line) over the past 410,000 years, measured relative to current temperature, constructed from an Antarctic ice core. Carbon dioxide (gray dotted line) is from air bubbles trapped in the ice.
Source: adapted from Petit et al. [1999].

record shows much more fine-grained detail than the longer data-sets above, including the shape of ice-age cycles with relatively short, warm interglacials (lasting 10,000–30,000 years) separating long, cold ice ages (lasting ~100,000 years). The cooling into an ice age is slow, taking several tens of thousands of years, while the warming at the end of an ice age occurs faster, in about 10,000 years. Note also that ice ages are only 5–8°C colder than today – a seemingly small difference considering the Earth is essentially a different planet during an ice age, with glaciers several thousand feet thick covering much of North America, sea level 300 feet lower than today, and all of the other accompanying changes in the world's environment and ecosystems.

Finally, Figure 3.11 zooms in one last time to show average Northern-Hemisphere temperature over the past 1,000 years, based on multiple proxies and modern records. Once again, this figure shows short time-scale temperature variations that are not visible in the graphs covering longer time spans. The various sources differ, particularly before about year 1500, but all show a similar pattern. Temperatures were warm 1,000 years ago, during a period called the *medieval warm period*. There were then several centuries of gradual cooling, bottoming out in a period 200–300 years ago called the *little ice age*, followed by faster warming since the nineteenth century.

This vast and growing body of information about the Earth's past climate places the climate of the last century in context. We can say with high confidence that over long time-scales back to a hundred million years ago, the Earth has been both far warmer and far cooler than today. We can also say with high

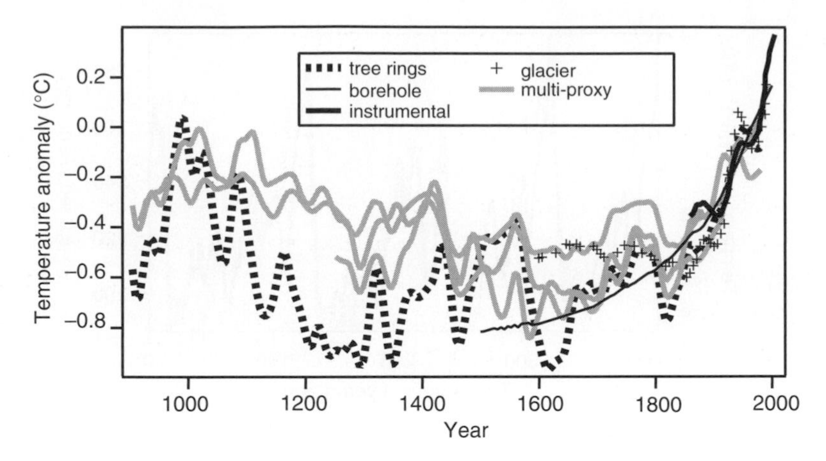

Figure 3.11 Average Northern Hemisphere temperature anomaly over the past 1000 years based on multiple proxy records and the modern surface thermometer record. Anomalies are calculated relative to the 1961–1990 average.
Source: adapted from Figure S-1 of National Research Council [2006].

confidence, however, that the last few decades of the twentieth century are warmer than any comparable period over the last 400 years, and possibly even warmer than the peak of the medieval warm period, around 1,000 years ago. We can also say that the warming of the last few decades has been rapid. For example, recent warming is occurring about 100 times faster than the average rate of warming that brought the Earth out of the last ice age.

In addition to providing context for the rapid recent warming, the relative climate stability of the past millennium raises the question of the robustness of human society to climate change. Rich, technologically advanced human societies have only developed over the past several centuries, and many anthropologists credit this success in part to the relatively warm and stable climate experienced over this period. We will consider impacts of projected climate change and societal vulnerability to changes in Section 3.4 and the next chapter.

In addition, long-term reconstructions of past climates deepen our knowledge of what conditions and what patterns of climate change are possible for the Earth's climate system, and provide a record to test theories of what factors drive climate variation and how sensitive the climate system is to them. Understanding the causes of these past changes can consequently help to inform the questions of how much of the change now occurring is being driven by human activities rather than natural processes, and how much more climate change we are likely to experience over coming decades, questions we examine in Sections 3.2 and 3.3 below.

3.1.8 Summary: is the climate changing?

Table 3.1 summarizes the information we have reviewed about trends in the Earth's temperature. All this evidence has been peer reviewed and multiply verified by independent scientific groups. No data set is perfectly reliable, of course. It is still possible that any one of these data sets could be significantly in error, although the critical scrutiny and multiple verifications each has received make this risk reasonably small. But there is essentially no chance that enough of these sources could be wrong by far enough, and all in the same direction, that the overall conclusion of substantial global warming in the twentieth century could be wrong.

Moreover, the data sources we have reviewed are just a small part of the mountain of evidence that the Earth is warming. Other corroborating evidence includes decreased Northern Hemisphere snow cover, thawing of Arctic permafrost, strengthening of mid-latitude westerly winds, fewer extreme cold events and more extreme hot events, increased extreme precipitation events, shorter winter ice season on lakes, and thousands of observed biological and ecological changes consistent with warming (e.g., poleward expansion of species ranges, earlier spring flowering and insect emergence, etc.). Some contrary evidence does exist, such as the lack of a decline in Antarctic sea ice, but such data are rare, regionally limited, and vastly outnumbered by evidence of warming. Under the weight of this abundant, consistent, thoroughly checked evidence, the IPCC's Fourth Assessment Report used remarkably strong language in concluding that recent global warming is now *unequivocal*.

3.2 Are human activities responsible for the observed changes?

This section considers the causes of the recent warming documented in the last section: *are human activities responsible for the recent observed warming, or might it be caused by some natural process?* This is a harder question than whether the Earth is warming, because establishing a cause-and-effect relationship requires an inference that simply identifying a trend does not. Showing human causation requires both demonstrating that human emissions can account for the observed warming trends, and showing that other potential explanations cannot.

While human emissions are clearly a potential cause of recent warming, other potential causes must also be considered. The Earth has experienced large climate fluctuations throughout its history, as Section 3.1.7 discussed, but it is only over the past few centuries that human activities have expanded to the point where they have the potential to influence global-scale processes, including climate.

Table 3.1 *A summary of measurements of changes in the Earth's temperature*

Type of Data	Direction of 20th Century Change	Size of Change, Comments
Surface thermometer measurements	Warming	Average surface air temperature increased about 0.7°C (1.3°F) over the twentieth century, with the rate of warming in the last half of the century about twice the rate of the first half.
Glaciers	Warming	Glaciers have been receding worldwide for the last two centuries, with evidence of faster retreat in the twentieth century.
Sea level change	Warming	Sea level rose about 17 cm total over the twentieth century, with the rate of increase accelerating in recent decades.
Sea ice	Warming	The area of Arctic sea ice has decreased by 2.7 percent per decade over the past 30 years, with decreases in summertime minimum area of 7.4 percent per decade. Average thickness of Arctic sea ice has also decreased over this time.
Ocean temperature	Warming	The heat content of the top 700 m of the ocean has significantly increased over the past 50 years.
Satellite temperature measurements	Warming	Satellite measurements since 1979 show warming broadly consistent with surface warming.
Climate proxies	Warming	The Earth's temperature was higher during the last few decades of the twentieth century than during any comparable period during the last 400 years.

Since natural causes must account for earlier climate changes, they must be considered as potential contributors to the rapid recent changes as well.

Five types of natural processes are known to have significant effects on climate: tectonic processes, variation in the Earth's orbit, volcanic eruptions, variation in the energy output of the Sun, and internal variability of the climate system. This section examines each of these natural processes as well as human greenhouse-gas emissions, asking how much each factor might account for the observed warming of the past century, especially the rapid warming of the past few decades. We conclude that, for the last half of the twentieth century at least, human emissions of greenhouse gases very likely account for most of the warming.

3.2.1 Tectonic processes

Tectonic processes are the geological processes involving the Earth's crust that control the locations of continents and ocean basins. These processes can influence climate in several ways. Whether continents are located in the tropics or near the poles determines how much land area can be covered by snow and ice, which affects the reflectivity of the planet. Thus, a continent moving from low to high latitudes, or vice versa, can change the climate. For example, the slow poleward drift of continents is believed to have initiated an ice age about 250 million years ago, in the Paleozoic era.

The location of continents also influences winds and ocean currents, which regulate the global climate by transporting heat from the tropics to middle and high latitudes. Thus, as continents move, this poleward heat transport can change. For example, the Antarctic Peninsula and the southern tip of South America separated about 30 million years ago, opening the Drake Passage. This opening allowed winds and water to flow unhindered in a continuous path around Antarctica. This intense flow effectively shut off transport of warm water and air from the tropics to the South Polar region, causing a dramatic cooling of the Antarctic, and contributing to the formation of the Antarctic ice cap.

Tectonic activity also can affect the climate by changing atmospheric CO_2. Atmospheric CO_2 dissolves in rainwater to form carbonic acid, the same weak acid found in carbonated soft drinks. When this rain reacts with sedimentary rock, in a process called *chemical weathering*, the CO_2 is converted to calcium carbonate which runs off to the ocean and is eventually buried in ocean sediment. The power of this process was illustrated by the collision, about 40 million years ago, of the Indian subcontinent with the Asian continent, which formed the Himalayas and the adjacent Tibetan Plateau. (Collisions between continents happen slowly: this one is still going on today.) The prevailing winds brought heavy rainfall onto the vast expanse of newly exposed rock in these features, and the resultant chemical weathering drew down atmospheric CO_2 over a period of millions of years.

Could any of these tectonic changes be responsible for the warming of the last few decades? The answer is almost certainly no, because tectonic processes are much too slow. Tectonic processes move continents only a few centimeters per year, so it takes millions of years to make a big enough change to affect the climate. Climate changes over a few decades or even a few centuries are much too fast for tectonic processes to exert any influence on them.

3.2.2 Orbital variations

The Earth's orbit around the Sun is not a perfect, unchanging circle, but an ellipse whose shape and orientation change slowly over time, in three ways. First, the eccentricity of the ellipse (the ratio of the major to minor axis) slowly varies, completing a cycle about every 100,000 years. As the eccentricity varies, so does the average distance between the Earth and the Sun. Second, the time of year when the Earth is closest to the Sun varies. The Earth is now closest during Northern Hemisphere winter, but in 10,000 years it will be closest in Northern Hemisphere summer. Third, the tilt of the Earth's axis relative to the Sun, now about 23°, slowly oscillates between about 22° and 25° over a period of 40,000 years.

These small orbital variations all affect the climate. The change in Earth-Sun distance changes the total solar energy the Earth receives. The other two changes do not affect the total sunlight reaching Earth, but change its distribution over the year and over the Earth's surface. For example, variation in the Earth's tilt changes how much sunlight falls on the tropics relative to the polar regions. It is now widely agreed that these slow orbital variations trigger the cycling between ice ages and warm interglacial periods that the Earth has experienced over the past few million years, as shown in Figures 3.9 and 3.10. This conclusion is based on near-perfect agreement between the timing of the orbital variations and the transitions into and out of ice ages.

So if orbital changes drive the climate changes of the past few hundred thousand years, could they also be causing the warming of the past century? They almost certainly cannot, for the same reason that tectonic processes cannot. These orbital wobbles are so slow that it takes thousands of years or longer to make any significant change in the pattern of incoming sunlight. The warming of the past century has been much too fast to be caused by these slow orbital variations. The warming must be due to other causes.

3.2.3 Volcanoes

Volcanic eruptions can change the climate in two ways. Volcanoes emit large quantities of CO_2, and so can contribute to warming the climate through the greenhouse effect. In fact, over time-scales of millions of years, the atmospheric abundance of CO_2 is controlled by the balance between volcanic emissions and slow removal into the ocean through chemical weathering and biological activity. Volcanoes can also have a faster effect on the climate, by blowing dust, ash, and sulfur gases into the atmosphere. The dust and ash fall to Earth quickly, but the sulfur gases combine with water to form small, suspended aerosol droplets that block incoming sunlight and cool the Earth for several years after a major eruption.

In 1816, for example, after three major eruptions in three years, the USA and Europe experienced the "year without a summer," in which snow fell in Vermont in June and heavy summer frosts caused crop failures and widespread food shortages. When that summer was followed by a winter so cold that the mercury in thermometers froze (this happens at –40°C), many residents fled the northeast USA and moved south.

Could volcanic eruptions somehow account for the observed recent warming? They could if volcanoes were responsible for the increase in atmospheric CO_2 over the last two centuries, but they are not. The observed buildup of atmospheric CO_2 is confidently attributed to fossil-fuel combustion by many pieces of evidence. Most convincing, the increase in atmospheric CO_2 has closely mirrored the amount of CO_2 humans have added to the atmosphere – it seems highly unlikely a natural process would do that. In addition, the isotopic signature of carbon added to the atmosphere is consistent with fossil fuels and inconsistent with volcanoes.

A second way volcanoes might have caused recent warming would be by a gradual reduction in volcanic aerosols in the atmosphere over the past century. But because the effects of each eruption last only a few years, this would require a series of massive eruptions every few years, each one precisely calibrated in its size and timing to achieve the required smooth aerosol reduction. We can test this claim by looking at the available records of eruptions over the last century or two, and of the amount of volcanic effluent in the atmosphere over the last few decades. While these records do appear to account for some of the bumps and wiggles in the global temperature record shown in Figure 3.1, there is no sign of the sustained pattern of eruptions required to explain the multi-decadal warming trend. As a result, we can safely rule out volcanoes as a cause of the recent warming.

3.2.4 Solar variability

Because sunlight is the ultimate power source for our climate, any change in the amount of sunlight reaching the surface can change the climate.

In fact, it is by changing the amount of sunlight reaching the surface that orbital variations and volcanoes exert their influence on the climate. But the amount of sunlight reaching Earth can also vary due to changes in the energy output of the Sun itself. It turns out that the Sun does not shine with constant brightness, but flickers like an old light bulb (it is 5 billion years old, after all). We do not notice this flickering, because it is small (fractions of a percent) and occurs slowly, over periods of months, years, and possibly longer.

The Sun's energy output cannot reliably be measured from the Earth's surface, but it has been measured since the 1970s from satellites. Over this period, the only variation observed has been the well-known 11-year solar cycle, by which total solar energy output varies about 0.1 percent. Because of the enormous thermal inertia of the oceans, the climate does not react to such short-term variations. And there have been no other changes in the Sun's output over this period that could account for the Earth's rapid warming.

Another reason to discount solar variability as an explanation is that an increase in solar energy would warm the entire atmosphere. This is not what is happening. Rather, measurements from weather balloons and satellites show that the stratosphere (the region of the atmosphere beginning around 10 kilometers altitude) has cooled over the past few decades. This pattern of change is consistent with warming from greenhouse gases – which warm the surface and lower atmosphere while cooling the stratosphere – but is not consistent with warming from solar variation. Thus, we can conclude with high confidence that the rapid warming of the past few decades is not caused by the Sun.

The Sun's influence on earlier climate changes, however, is more difficult to determine. Solar output before the period of direct satellite measurements must be inferred indirectly from proxies, such as the number of sunspots, which people have counted for thousands of years. The most recent analyses of these records have concluded that the Sun has brightened over the last few hundred years, and this can potentially explain at least some of the gradual warming of the last few hundred years, but not the faster warming of the last few decades.

3.2.5 Internal variability

All the potential sources of warming discussed so far involve *forced variability*, changes in the Earth's climate in response to some externally imposed change, such as a change in the distribution of incoming sunlight, the reflectivity of the Earth, or the arrangement of continents. But the Earth's climate system is so complex that it can change without external factors driving it, rather like the wobbling of a spinning top. Such changes are called *internal variations*, of which the best-known example is the El Niño/Southern Oscillation discussed in Section 3.1.6.

Could the warming of the last few decades be internal variability, part of some natural oscillation of the climate system? The proxy data on climate variations before the past two centuries can help answer this question. Human activities probably had minimal impact on climate before 1800, so climate proxy data before that time should provide a good picture of recent natural climate variability. But as Figure 3.11 shows, the record between AD 1000 and 1800 shows nothing similar to the rate and magnitude of warming of the twentieth century. So if recent warming is due to natural variability, there is no evidence it is a type of variability that has been operating over the past 1000 years.

As one looks at older proxy data, the time resolution of the data declines, so the ability to see decadal or century-scale variations disappears. Thus, we know less about how temperature varied from year to year as we go further back in time. There is some evidence of rapid climate changes over the past twenty thousand years, with warming of up to a few degrees Celsius occurring over a few decades to a century, during transitions into or out of ice ages. But these rapid natural changes occurred with rapid, large-scale reorganization of circulation patterns in the atmosphere and oceans, and there is no sign of similar large-scale circulation changes occurring today.

An additional problem for explaining recent warming as natural variability is that the spatial distribution of the observed warming appears inconsistent with internal variability. Any form of internal variability responsible for the recent warming would almost certainly be driven, like El Niño, by changes in ocean circulation that warm the ocean surface. If this were happening, we would expect to see more warming over ocean than over land, but observations show the opposite: land areas have been warming significantly faster than the oceans.

Finally, we can gain insight into natural climate variability by running climate models without any human greenhouse-gas emissions. When used this way, climate models exhibit variations in global-average temperature from year to year and decade to decade that are similar to those seen in the climate proxy data before about 1800, but they produce nothing resembling the rapid warming of the past century. (We will see in the next section that only when recent increases in greenhouse gases are included can climate models generate the observed rapid recent warming.) Considered together, all this evidence suggests that while we cannot definitively exclude natural climate variability as a contributor to recent warming, it is unlikely that it can account for any significant fraction of the recent rapid warming.

3.2.6 Increases in greenhouse gases

Section 1.2 discussed how greenhouse gases have been increasing in the atmosphere for the last two centuries or so, primarily as a result of human activity. There are strong theoretical reasons, rooted in basic physics, to expect this increase to warm the Earth's surface. The past climate record shows many examples of temperature fluctuations associated with changes in CO_2 and other greenhouse gases, providing support to this basic theoretical expectation. Figure 3.8, for example, shows warm temperatures 50 million years ago when CO_2 was several times higher than today, followed by a long slow cooling in parallel with decreased CO_2 from increased chemical weathering and biological ocean uptake.

In addition, the sharp spike in temperature about 55 million years ago in Figure 3.8, the Paleocene-Eocene thermal maximum or PETM, vividly illustrates the power of greenhouse gases to warm the planet. At that time, a massive release of CO_2 and methane into the atmosphere over about 10,000 years led to an abrupt warming of about 5°C. As the greenhouse gases were removed from the atmosphere over the next 100,000 years, the Earth's temperature returned to its pre-PETM value.

The association between CO_2 and temperature is even clearer over the last few hundred thousand years. Figure 3.10 shows how CO_2 and temperature varied in lock-step as the Earth cycled between ice ages and warm interglacials, but the dynamics of this relationship were somewhat complex. We know that ice ages are initiated by small variations in the Earth's orbit, but the resultant changes in sunlight distribution are too small to directly account for the large changes in global temperature that follow – something else must be magnifying the influence of the orbital variations. Most scientists now think this is a positive feedback by which a small initial warming triggers a release of CO_2 and other greenhouse gases, which cause further warming. The precise mechanism of this feedback is not well understood, but one possibility being considered is that the initial warming increases biological activity, which releases CO_2 from increased decay of organic soils.

The final piece of evidence that greenhouse gases are responsible for recent warming comes from climate models. Figure 3.12 compares the observed temperature record since 1900 with results of two climate model runs. The model run in panel (a) includes the known natural forcings – solar variability and volcanoes – but no human impact on climate. This calculation reproduces many of the bumps and wiggles in the record, suggesting that these are not due to human activity. But this simulation completely fails to capture the rapid warming beginning around 1970.

The model run in panel (b) includes both natural effects as well as the effects of human activities – mainly greenhouse gas emissions, but also including sulfate aerosols and stratospheric ozone depletion, both of which cool the surface.

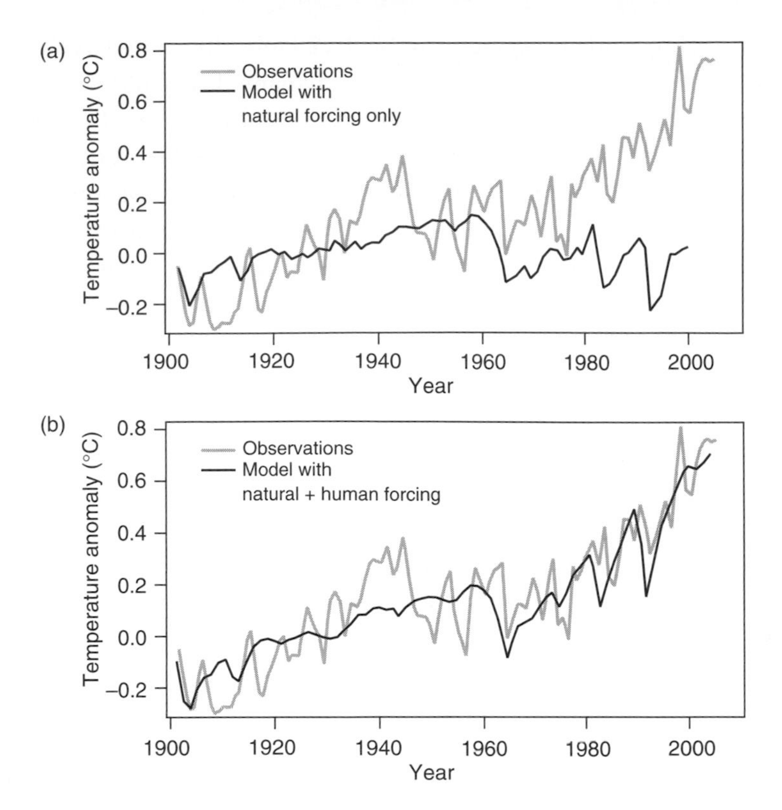

Figure 3.12 Global mean surface temperature anomalies from the surface thermometer record (gray line on both plots), compared with a coupled ocean-atmosphere climate model (black line). (a) Model includes only non-human natural climate forcing, in particular solar and volcanic effects. (b) Model includes both natural forcing and human greenhouse-gas emissions, aerosols, and ozone depletion. Anomalies are measured relative to the 1901–1950 mean.
Source: Fig. TS.23, IPCC [2007a].

This model captures most of the important features in the observations – in particular it captures the rapid warming since 1970 that the model with only natural forcing fails to simulate. This shows that human greenhouse-gas emissions, volcanic, and solar effects have all contributed to global temperature changes of the past century, but that greenhouse-gas emissions are responsible for most of the rapid late twentieth-century warming.

3.2.7 Summary: are human activities responsible for recent warming?

We have considered six potential causes for the observed warming of the last few decades of the twentieth century. Two of these, orbital variations and tectonic processes, can be decisively eliminated as significant contributors because they are too slow to have any discernible effect on climate over periods as short as a few decades or a century. Two other processes, volcanic eruptions and changes in solar output, can also be rejected, because we have good measurements of them over the relevant period and they have not shown the pattern of changes that would be required to account for the recent warming. Internal variability probably accounts for many of the observed wobbles and variations in climate over the past century or two, but there are also strong reasons to reject it as the cause of the sharp warming of recent decades.

This leaves only the increase in greenhouse gases, and we have described the evidence supporting this explanation. Section 1.2 outlined the strong theoretical reasons why we would expect increases in greenhouse gases to warm the planet. In addition, greenhouse gases and temperature have been linked numerous times over the last 60 million years of climate history. In fact, the evidence connecting greenhouse gases to climate is so strong that to claim the rapid recent warming has some other cause, it would be necessary to explain why the observed increase in greenhouse gases is *not* warming the planet as we would expect it to. The capstone evidence is provided by climate models, which can reproduce the climate changes of the past century only if they include greenhouse gas emissions from human activities.

Given the compelling evidence supporting greenhouse gases, and the lack of any plausible alternative explanation – despite many attempts to find such evidence – the IPCC concluded in its 2007 report that "Most of the observed increase in global average temperatures since the mid-twentieth century is very likely due to the observed increase in anthropogenic greenhouse gas concentrations."

This conclusion is very strong, but it is also important to note the three ways in which it is carefully limited. First, the Earth has been warming for at least 400 years. For warming before the middle of the twentieth century, however, a significant contribution from natural factors such as solar variability cannot be ruled out. Human activities may well have contributed to this earlier warming. But this has not been demonstrated with the required high confidence, so the IPCC statement specifically avoids attributing pre-1950 warming to human activities.

Second, the IPCC does not say that natural factors such as internal variability made no contribution to the warming of the last few decades. In fact, natural factors probably did exercise some influence over this period. What the evidence supports is that any contribution from natural factors is small relative to human influences over this period. Consequently, the IPCC reports say that humans are responsible for "most" of the recent warming, not all of it. Finally, while we know much about the climate system, some uncertainties remain. For this reason, the IPCC describes this conclusion as "very likely," meaning in their carefully nuanced language they judge it 90% likely to be true.

How else do people modify the climate?

Greenhouse gases are just one of several ways that human activities can change the climate. Human activities also increase the atmospheric abundance of aerosols – tiny particles, either solid or liquid, suspended in the atmosphere, which can either warm or cool the Earth's surface depending on their composition.

Inefficient or incomplete combustion, as occurs in two-stroke engines and in low-temperature cooking fires burning traditional fuels such as dung or crop residues, releases black carbon aerosols (tiny particles of soot), which absorb both incoming sunlight and upwelling infrared radiation, thereby warming the surface. Burning of fuels containing sulfur forms liquid sulfate aerosols, which reflect incoming solar radiation back to space and so cool the surface. Aerosols also interact with clouds, increasing their reflectivity and thereby cooling the surface.

Changes in land-use can also affect the climate. Cutting down a forest and replacing it with farmland, for example, replaces a dark vegetation surface with a lighter one, reflecting more sunlight and cooling the climate.

The dominant climate effect of land-use changes is local, but these can be significant globally when the scale of land-use change is global. Adding up all the non-greenhouse gas climate effects of human activities, they are causing a net global cooling that is probably offsetting about 30% of the warming from greenhouse gases, with most of this cooling coming from reflective sulfate aerosols. Since they are a major component of air pollution, efforts to reduce air pollution will reduce this offsetting cooling effect and cause additional warming.

3.3 What further changes are likely? Predicting twenty-first-century climate change

Identifying past climate changes and the extent to which emissions from human activities have been responsible are important achievements in climate science. But it is the threat of future climate change that drives public concern and policy-making. Making informed decisions of what to do requires information about what climate changes we might face in the future, and how our actions can moderate them. This need puts predictions of future climate change at the center of the policy debate.

Climate models are the primary tool for predicting future climate change. We have already seen how climate models have played a key role in attributing the observed recent warming to greenhouse gases. In those calculations, climate models use the observed atmospheric concentrations of CO_2 and other greenhouse gases as inputs to simulate the climate of the last century. Climate models can also be used to project future changes, but using them this way requires predictions of future concentrations of greenhouse gases as inputs. Projecting future atmospheric concentrations requires projecting how much CO_2 and other greenhouse gases will be emitted from human activities.

Emissions projections are not a matter of atmospheric science, but an exercise in predicting societal trends. Future emissions depend on future world population and economic growth, because more people and more economic activity mean more energy use, industry, agricultural production, and other activities that generate emissions. Future emissions will also depend on policies, and on technological trends that determine the efficiency of energy use and the mix of carbon-emitting and non-emitting energy sources in use. Historical experience, as well as economic, social, and demographic research, provides some insight into future trends, particularly to exclude some ranges as implausible (e.g., a sudden cessation of population growth in a young population, or an extended period that combines stagnant economic growth with rapid technological innovation). But our knowledge of social and economic dynamics is not sufficient to allow a single confident prediction of future emissions.

Consequently, the approach taken by the IPCC is to produce a set of emission scenarios, where each scenario provides an alternative, internally consistent, plausible picture of how world development might shape future emission trends. Together, the set of scenarios roughly spans the range of alternative emission futures judged plausible. (Of course, a set of scenarios cannot cover all possible emissions trends, since these will also depend on large-scale historical events such as wars or political transformations that we have little ability to predict or control.)

The IPCC has supported several exercises to develop scenarios of future emissions, which have been used to provide consistent emission inputs to drive climate-model projections. A major scenario exercise in 1992 developed five

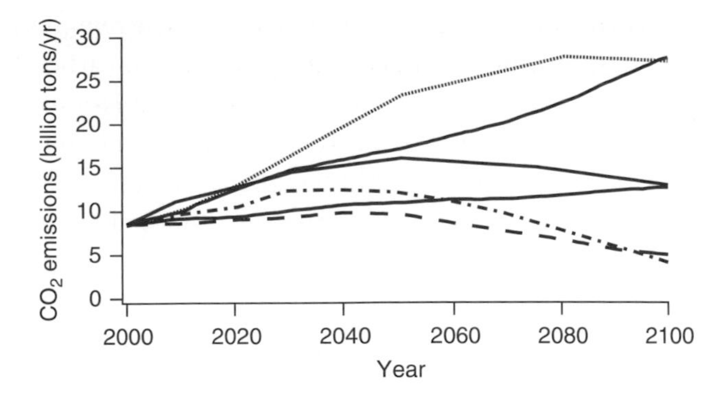

Figure 3.13 CO_2 emissions, in billions of tons carbon per year, from the IPCC scenarios in the Special Report on Emissions Scenarios (SRES). The dotted line is the scenario A1FI, the dashed line is the scenario B1, the dot-dashed line is scenario A1T. *Source*: Fig. 17 of the Technical Summary, IPCC [2001a].

scenarios, including a middle "reference case" under which world emissions grew from their initial value of 8 billion metric tons, or gigatons, of carbon-equivalent (GtCe) to about 14 billion in 2050 and 20 billion in 2100.[2]

A second scenario exercise in the late 1990s avoided defining a central case, but rather sketched four alternative pathways of world development, and a total of six 'marker scenarios' to serve as benchmarks for climate-model projections: one scenario from each development pathway, plus two variations of energy technology assumptions within one development pathway. All these scenarios assumed that no concerted actions to reduce emissions were adopted. Figure 3.13 shows projected carbon dioxide emissions for these six marker scenarios through 2100. These emissions scenarios were used to drive model projections of future climate change in both the third and fourth assessment reports of the IPCC, published in 2001 and 2007. As Chapter 4 will discuss, a new set of scenarios is now in preparation that emphasizes alternative pathways to global climate stabilization.

Projected emissions in these scenarios span a wide range, from as high as 30 billion tons to as low as 5 billion in 2100. This wide range reflects combined uncertainties about population, economic growth, and technological trends. For example, the dotted line, which shows emissions reaching 30 billion tons by 2100, assumes a moderately optimistic continuation of recent trends – relatively low population growth, high economic growth, and gradual convergence of

[2] Units are in metric tons (also called tonnes), equal to 1,000 kg or about 2,200 lbs. Expressing total greenhouse-gas emissions as gigatonnes of carbon-equivalent means that non-CO_2 gases are converted to a quantity of CO_2 with the same effect on radiative forcing. This CO_2 is then measured by the mass of the carbon, excluding the mass of the oxygen atoms in the CO_2 molecule.

incomes between world regions. It also assumes that energy use remains heavily dominated by fossil fuels, shifting toward coal and other higher-carbon fuels as low-cost oil and gas decline.[3] The dashed line, which shows emissions peaking below 10 billion tons in mid-century then declining to about 5 billion by 2100, represents an optimistic vision of what might be called a "sustainable development" future. It projects the same population growth as in the high-emissions scenario, somewhat slower economic growth but with lower energy and material intensity due to a shift of the world economy from manufacturing to services and information, as well as rapid adoption of low-carbon energy technologies.

The dot-dashed line assumes the same population and economic growth as the high-emissions scenario, but sharply different technological trends. Instead of a shift toward higher-carbon fossil fuels like coal, this scenario assumes technological developments that shift energy supply toward non-emitting sources. Emissions in this scenario rise until mid-century when they peak between 10 and 15 billion tons, then decline even further than those in the scenario above – illustrating the powerful effect of technological developments in determining emissions. Chapter 4 provides more detail about scenarios, the assumptions underlying them, and what they imply for possible actions.

A particular climate model, using a particular emissions scenario as the input, will generate a projection of future climate change. But there are about 20 state-of-the-art climate models in use, each developed by a different scientific group, which differ in their approaches to simulating the atmosphere. They may break the atmosphere up into different size boxes, give greater or lesser emphasis to certain atmospheric processes, or use different computational approaches to represent basic climatic processes, particularly those that must be parameterized because they operate at scales too fine to represent explicitly. Because of these differences, different models project different climate futures, even when driven by the same emissions scenario.

Figure 3.14 summarizes model projections of climate change over the twenty-first century by a range of climate models for the emissions scenarios. The left panel shows evolution of world temperature through 2100 for the median climate-model projection for three of the six marker scenarios. It also shows, in line C, the median projection of future warming if emissions suddenly ceased in 2000. This shows the additional warming already committed by past emissions because of lags in the climate system, about 0.4°C above the level of 2008.

[3] This scenario is named A1FI. A1 labels the basic development pathway, while "FI" stands for "fossil-fuel intensive." The low-emissions scenario described below is called B1, while the scenario with the shift to non-fossil sources is called A1T ("T" is for advanced technology).

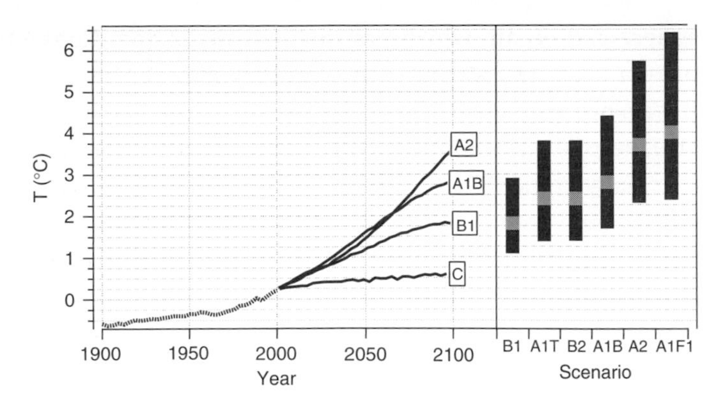

Figure 3.14 Left: Average surface-temperature anomaly predicted by climate models for the twenty-first century, relative to the 1980–1999 average. Dotted line shows measured temperature anomaly over the twentieth century, as in Figure 3.1. Line C fixes greenhouse gases at year 2000 values; B1, A1B, and A2 scenarios represent low, medium, and medium-high greenhouse gas emission scenarios. Right panel: Temperature anomaly ranges for the six marker scenarios in the year 2100. For each scenario, the gray region is the best estimate, while the black bars show the likely range.
Source: Figure SPM.5 of IPCC (2007a).

The right panel shows the range of model-calculated temperatures in 2100 for each of the six emission scenarios, with the median or "best estimate" climate projection for each scenario highlighted in the middle of the vertical band. Considering all emissions scenarios, these best-estimate projections range from 1.8 to 4.0°C. Considering uncertainty in both emissions and climate-system response, projected twenty-first-century warming ranges from 1.1 to 6.4°C. To put this in context, even the lowest projected warming, which combines the lowest emissions scenario with the lowest climate sensitivity, is significantly larger than the 0.7°C warming experienced in the twentieth century. The largest warming, 6.4°C, is similar to the warming since the last ice age.

This is a sobering picture. While there is a wide range of uncertainty in the magnitude of future warming, warming is projected to continue through the twenty-first century by all models under all emissions scenarios. If future warming falls in the middle of the range in Figure 3.14, which we must assume is a more likely outcome, or near the top of the range, then the rate of warming over the century would be extreme. The only historical evidence of global temperature changes as large and as fast as the high projections for this century is a series of abrupt warmings and coolings that occurred at the end of the last ice age. But these changes accompanied major reorganizations of the circulation of the atmosphere and ocean, which are not now occurring. Consequently, the upper range of projected warming this century may represent a climate change that has no precedent over the entire history of the Earth.

To summarize, despite uncertainty in both emission projections and climate models, it is virtually certain that the Earth's temperature will continue to increase. We are committed to some warming – even if we stopped emitting greenhouse gases today, the Earth's temperature would warm several tenths of a degree Celsius over the next few decades. Given the likely evolution of emissions over the next century, the global average temperature by late this century will likely be about two to four degrees Celsius warmer than the present, possibly higher – unless emissions are reduced sharply.

3.4 What will the impacts of climate change be?

Change in global-average temperature is a yardstick used to measure climate change, but is not what people care about. Climate change matters because of the resultant changes where people live, in local climate and weather and its effects on people and things they value. Describing climate impacts consequently requires projecting climate change for specific regions and seasons, where people and climate-sensitive systems actually experience the climate. It also requires projecting not just temperature, but also other characteristics of climate, especially precipitation. And it requires projecting not just changes in annual average values, but also changes in their seasonal cycle, variability, and extremes.

These requirements pose severe challenges to climate modeling and projection. As projections move from the global average toward smaller regions, they benefit less from cancellation of smaller-scale errors, so forecast errors grow larger. It is especially difficult to project precipitation at regional scale because it can vary greatly over short distances. Small changes in storm tracks can radically shift the location and seasonal distribution of precipitation. Projecting climate impacts also requires estimating the responses of climate-sensitive ecosystems and resources, introducing further uncertainty to projections. Moreover, in many areas of likely climate impact such as agriculture and commercial forests, human management dominates the systems, so assessing climate impacts requires considering human responses to the changing climate. We discuss this last issue, and other socio-economic aspects of impacts and adaptation, in the next chapter. Here we focus on direct bio physical impacts of climate change that can be projected without having to consider human responses.

Despite all the difficulties, we do know some things about the impacts of climate change. We know that some impacts are harmful, others are beneficial, and many are mixed – harming some people, places, or activities, and benefiting others. Most analyses of impacts suggest that harmful impacts are likely to outweigh beneficial ones, by a little in places that are rich, well governed, and adaptable, and by a lot in places that are less fortunate. If climate changes are large or happen quickly, harmful impacts are increasingly likely to dominate beneficial ones, even in rich and well-governed regions. Overall, knowledge varies greatly across types of impact: some are understood fairly well, others poorly. And there may be complex connections among climate-sensitive systems that create vulnerabilities to climate change that have not yet been recognized.

Where understanding of impacts is good, it is often because regional or even local impacts are strongly coupled to global changes by well-known physical processes. A particularly clear example is sea level rise. A warming climate will raise sea level, through thermal expansion of seawater and melting of glaciers. The range of global warming shown in Figure 3.14 is projected to bring further sea level rise of 18–59 cm by 2100. Since a half-meter rise in global sea level, roughly speaking, means a half-meter rise on every coastline, what this will mean for any specific location can be readily, if approximately, assessed. How serious it will be in each place will depend on local factors, such as the amount of low-lying coastal land, settlement and land-use patterns, property values, and the resources available to manage an appropriate mix of coastal protection and orderly retreat.

This range of sea level rise projections includes no contribution from potential loss of major ice sheets in Greenland and West Antarctica. These ice sheets, thousands of feet thick, each contain enough water to raise global sea level by 4–6 meters (about 13–20 feet). It is known that during the last warm interglacial period sea level was 4 to 6 meters higher than today, suggesting large-scale loss of ice from one or both of these sheets. Since projected warming this century will match or exceed the temperature of the last interglacial, this warming will probably raise sea level by several meters – *eventually.* The key uncertainty, however, is how fast this will happen, because the dynamic response of these ice sheets to warming is not well understood. The most benign possibility would be that they slowly melt in place, like giant ice cubes. In this case, given the large thermal inertia of ice, it might take a millennium or longer for the glaciers to fully melt. Alternatively, the sheets could be profoundly destabilized by warming. One possible mechanism is melt water, which forms on the top of the glacier during the summer, can melt through the ice sheet and reach the bedrock. There, it warms and lubricates the ice-rock interface and allows the glacier to flow more rapidly into the ocean. There are recent – and controversial – suggestions that such processes could result in sea level rise of as much as a few meters this century. Given current low confidence about the mechanisms and rate of loss of these ice sheets, the IPCC simply excluded their potential contributions in its projections of twenty-first century sea level rise.

Another impact strongly linked to global-scale changes comes from elevated atmospheric CO_2 directly, rather than its effect on climate. About half the CO_2 emitted to the atmosphere is rapidly absorbed by the ocean, where it is converted to carbonic acid, the same weak acid found in carbonated drinks. As atmospheric CO_2 continues to increase, it is virtually certain that the oceans will become more acidic. This may have severe implications for ocean ecosystems, through hindering formation of the calcium carbonate shells and skeletons on which many marine organisms, including corals, depend.

Most other impacts require examining climate projections at relatively fine regional scale, with consideration of local topography, ecosystems, and current climate. The implications of a few degrees warming may be very different depending on whether it occurs in a forest or a desert. Although climate projections grow more uncertain at smaller regional scales, some broad regional results are now well established, because they appear consistently across many climate-model projections and are grounded in basic physical principles. For example, it is virtually certain that the continents will warm more than the oceans, because of the moderating effect of the ocean's huge heat capacity. Climate models project that northern North America and Eurasia will warm more than 40 percent more than the global average. Other confident projections include that most land areas will experience more frequent extreme hot days and nights and heat waves; that warming will be larger at night than in the day and larger in winter than in summer, so both daily and annual temperature ranges will decrease; and that warming will be greater at middle and high latitudes, particularly in the Northern Hemisphere, than in the tropics.

In fact, Arctic and sub-Arctic regions are already experiencing extreme warming, with severe impacts on many resources and activities. Thawing of permafrost, retreat and thinning of sea ice with resultant increases in coastal erosion and disruption of marine ecosystems, and shorter ice-travel seasons on lakes and rivers, have already brought large disruptions. These are likely to accelerate under the large further warming projected for the Arctic this century. All climate models project continued retreat of Arctic sea ice through the century, with the Arctic becoming completely ice-free in the summer possibly within a few decades. Loss of summer Arctic sea ice would have implications for global ocean circulation and climate that are potentially enormous, although not yet well understood.

In addition, an even partly navigable Arctic Ocean would have huge effects on shipping, Arctic development, and military operations and security. And this may be occurring even faster than the models project. During the summer of 2007, enough sea ice melted to open the "Northwest passage" through Canada's Arctic islands. Normally passable only to ice breakers, this passage is a much shorter route from the Atlantic to the Pacific, or from the east to west coast of North America, than sailing through the Panama Canal or around Cape Horn. Given the near certainty of further Arctic warming, this passage will very likely keep opening, and soon may do so every year. Loss of Arctic sea ice may also trigger to a rush to exploit previously inaccessible natural resources, including large oil and gas reserves, carrying significant potential for international conflict. In 2007, Russia made a claim to Arctic resources by planting a flag on the ocean floor near the North Pole, a claim that is disputed by both Canada and Denmark.

In mid-latitudes, continental regions are projected to experience substantial warming (e.g., 3–6°C over the continental United States), although projections of how this average warming will be distributed across continents are quite variable. Higher summer temperatures and higher humidity together will bring substantial increases in the summer heat index, a measure that combines heat and humidity to estimate how hot it feels. Some models project an increase of up to 5–14°C in July heat index in the southeastern USA. If you live there, you know how miserable that would be.

Despite the difficulty of simulating precipitation in climate models, some precipitation changes are projected with confidence. It is very likely that precipitation will increase at high latitudes and decrease in the subtropics (latitudes of 20° to 30°, where the world's great deserts lie), continuing recent trends. Also continuing a twentieth-century trend, more precipitation is likely to come in the heaviest downpours, bringing increased erosion and higher risk of flooding and landslides. When rain falls in heavy downpours, more of it runs off and less is absorbed by soil or stored in reservoirs. Combined with warmer summer temperatures, which will increase water loss from soils through evaporation, this leads to the surprising projection that both wet and dry extremes will grow more likely: wet extremes, with associated risks of flooding, increased erosion, and landslide; and dry extremes, with associated risks of water shortages, crop loss, wildfire, and increased vulnerability of crops and forests to pests and disease.

In some cases, the broad implications of these direct biophysical impacts of global climate change to human affairs may be obvious. In other cases, understanding them requires detailed analysis of the behavior of climate-sensitive systems. For example, it is obvious that changes in the amount, location, and timing of precipitation can alter freshwater availability, but quantitative projections of freshwater changes and their implications require detailed study of specific water systems, including how people manage them.

A study of climate impacts on the Columbia River in the US Pacific Northwest provides an important example of the kind of analysis required. While projected changes this century in total annual precipitation and Columbia streamflow are small, projected warmer, wetter winters and hotter, drier summers are likely to shift the seasonal pattern of streamflow. Because the Columbia draws much of its flow from melting of accumulated winter snowpack, its flow now peaks in late spring. But under projected warmer winters, more winter precipitation will fall as rain rather than snow. This will increase flow in winter, when water is already abundant in the region, and decrease it in summer, when the region is already acutely water-scarce. Similar changes are likely in other regions that meet water needs in dry summers by drawing on snow-fed rivers, highlighting the importance of examining not just total annual water availability, but details of seasonal flows.

Climate change will unquestionably affect natural or unmanaged ecosystems. The distributions of plant, animal, and microbial species are influenced by many factors, but climate is a major determinant. Changed climate will affect many aspects of the reproduction, behavior, and viability of species in diverse ways, thereby changing species' spatial range and inter-species relationships. There is abundant evidence that these changes are already underway in response to recent climate change, including shifts of species ranges poleward and to higher elevations, and changes in the timing of seasonal events such as tree leafing, leaf-fall, and egg-laying.

Under continuing climate change, present ecosystems will not simply move intact to follow their optimal climate to new locations. Rather, the individual species in the ecosystem will be affected in particular ways. Some may migrate readily, while others may be unable to move fast enough to keep pace with the changing climate. Species ranges will adjust at different rates and by different processes, in many cases subject to other human interventions and constraints such as land-use change, barriers, and intentional or inadvertent transport.

The aggregate result will be that present ecosystems are continually disrupted and reorganized, with new relationships among incumbents and new arrivals continually re-established in each location. In some cases, the new assemblies may be similar enough to present systems that thinking of present ecosystems simply being shifted (for example mixed-temperate forests shifting north into the present boreal forest zone, boreal forests shifting north into the present tundra zone) is not too misleading. In other cases, however, the new systems may be unlike present ecosystems, exhibiting novel species relationships, extinction of species with narrow geographic or climatic distributions, or other ecological surprises. The consequences for ecosystem services such as water retention and nutrient cycling, and for ecosystem amenities such as opportunities for human uses and recreation, are likely to be substantial. Some ecosystem types are likely to be lost entirely, because of physical limits or barriers to the movement of key species, or the complete loss of the required climate conditions. In the United States, ecosystems threatened with near or total loss include alpine systems in the lower 48 states, coastal mangrove systems and coral reefs.

A particularly important factor in ecosystem impacts will be the rate of climate change. Ecosystems have adapted to climatic variations in the past, but past changes have generally been much slower than those projected for the coming century. It is virtually certain that ecosystems will adapt less easily to the predicted rapid changes than to the slow changes of the past few thousand years. What is uncertain is how much less easily, and with what consequences.

Systems managed for human use, such as agriculture, commercial forests, rangelands, and aquatic and marine systems (fisheries, etc.), are also sensitive to climate and related changes, but are dominated by human management. This has two implications for projecting climate-change impacts, which work in opposite directions. On the one hand, disruption of these systems by climate change may have severe human impacts because we depend on them so much. On the other hand, the ability to adapt management practices to changing conditions offers the possibility of mitigating these harmful impacts. We discuss the linked issues of climate impacts and adaptation in the next chapter.

We can roughly summarize present knowledge about the impacts of climate change as follows. For rich, mid-latitude countries such as the USA, Europe, and Japan, impacts of climate change this century might range from small to severe. These countries, however, have substantial financial, technological, managerial, and political capacity to adapt to harmful impacts – unless climate change lies near the top of the projected range shown in Figure 3.14, in which case even these countries are likely to face serious challenges. Poorer countries, mostly located in the tropics and sub-tropics, are projected to face climate changes different in detail but at least as challenging as those projected for the mid-latitudes. Because these countries have fewer resources to adapt to impacts, the consequences for them may be severe even for climate change near the lower end of the projected range.

Moreover, it is crucial to note that, while most of the plots in Section 3.3 stop in 2100, climate change does not. Under all uncontrolled emissions scenarios, atmospheric CO_2 and temperatures continue to rise well beyond 2100. Although uncertainties in climate projections grow larger as we look further in the future, the analyses that have looked beyond 2100 suggest that climate change and impacts will grow increasingly severe. There is growing evidence that benefits to plant growth from elevated CO_2 level off over time as CO_2 continues to rise, while stresses from climate change continue to increase. Consequently, unless there is an extreme level of technological and economic progress that frees us from dependence on anything resembling crops in fields or relatively natural forests – which may well happen, since 100 years can bring vast economic and technological changes – impacts of unabated climate change beyond 2100 look increasingly serious and potentially unmanageable, even for the rich countries of the world.

Finally, it is also necessary to consider the possibility of climate surprises: high consequence, possibly sudden changes that either appear to be quite unlikely (but which cannot be ruled out), or that we may completely fail to anticipate. One example of such a potential extreme event, as discussed above, would be the rapid loss of a major ice sheet in Greenland or West Antarctica, raising global sea levels several meters over the century. The flooding of coastal areas worldwide that would result would represent an unimaginable environmental and humanitarian catastrophe. Other potential extreme events that have been proposed include large-scale reorganization of ocean circulation, or some large positive feedback linking warming with changes in the global carbon cycle. For all these cases, most relevant experts currently judge the events unlikely to happen this century, but their risks are not well understood and are clearly not negligible.

3.5 Contrary claims

This chapter has summarized the state of knowledge on the key scientific questions that underpin concern about climate change, including the evidence that the Earth is warming, that greenhouse-gas emissions from human activities are the predominant cause, that warming will continue over the next century, and that despite uncertainty in the rate and regional details of future climate change, there is significant risk of serious, possibly severe impacts. We close the chapter by noting that there are widely publicized statements made in climate policy debates that deny all these points. These statements are sometimes made by political actors who oppose action on climate change, sometimes by people with scientific credentials who claim to be expressing scientific criticisms, who are often called "climate-change skeptics" or "climate-change deniers." They assert not just that the prevailing view of climate change summarized above is wrong, but also that the supposed scientific consensus on these points does not actually exist, but is an attempt to suppress legitimate scientific dissent in order to advance an activist political agenda.

In this section, we review a few of the most prominent of these claims and discuss why they are confidently believed to be wrong. As one reads this section, two cautions are in order. First, we are not arguing that current knowledge of climate is perfect. There is plenty of uncertainty and dissent in climate science, as in any active field of science. Indeed, this is where one finds the interesting questions that scientists pay attention to. But there are also many points of climate science that are known with great confidence. Second, we are not suggesting that it is irresponsible or dishonest to question scientific knowledge with high stakes for public welfare and action, even on points that appear to be thoroughly settled. Highly critical commentary can be valuable in identifying weaknesses in current understanding – when based on relevant knowledge and using scientific standards of argument – and even outlandish proposals occasionally turn out to be correct (although they usually do not).

But the main claims of climate skeptics are not, for the most part, being advanced in a manner or setting that offers any prospect of contributing to the debate, illuminating untested assumptions, or advancing knowledge. They are usually not advanced in scientific arenas at all, but in newspaper opinion articles, on the Internet, or in other outlets where standards of evidence and argument are lenient and there is no peer review. They can appear persuasive, both because they sound plausible to those unfamiliar with the scientific debate and because they are often coupled with broader political arguments or powerful rhetorical devices such as inflammatory rhetoric or ad hominem attacks. In a few cases these have advanced criticisms that on examination have turned out to be correct, but unimportant. In others, they have used biased data such as selective start and end dates for trend claims, or advanced claims that are unsupported, eccentric, previously rejected, or transparently false. The media's disposition to uncritically balance opposing views has often given these marginal or clearly erroneous claims the same stature as well supported consensus scientific views.

There are many of these claims in circulation, and they change over time. Here, we summarize a few of the most prominent that have been advanced as scientific claims within the past several years, and explain why they are confidently believed (or in some cases certain) to be wrong.

Claim 1: the Earth is not warming, or global warming has stopped.

Claims denying the evidence that the Earth is warming have circulated more or less continuously for decades, but their precise form has evolved over time as evidence for warming has mounted. Until a few years ago, some advocates simply denied that the Earth had experienced any warming at all. This claim was sometimes supported by analysis of small regions where cooling was actually occurring. These are not valid as guides to global trends, however, because trends in small regions can differ greatly from global trends, even going in the opposite direction. The claim was also frequently supported by early analyses of satellite data, published in the mid-1990s, which showed a global cooling trend in the lower atmosphere. But further examination of the satellite trend revealed several serious errors in these early results, such as failing to correct for changes in the satellite's orbit. Correction of these errors, plus the accumulation of more years of data, have brought the satellite and surface records into reasonable agreement, with both showing warming, as discussed in Section 3.1.6.

The most recent variant of this claim acknowledges that the Earth has warmed in the past century, but says that warming stopped in 1998 – or even, in extreme cases, that since 1998 the Earth has entered a new cooling period and may be headed for another ice age. It is correct that 1998 was an exceptionally hot year, mainly due to the biggest El Niño of the twentieth century. No year since 1998 has been hotter, although 2005 equaled it.[4]

But even if no year since had equaled the temperature of 1998, this would not mean that "global warming" had stopped. In addition to the long-term warming trend underway, the global-average temperature fluctuates from year to year, due to El Niño and other forms of climate variability. In recent years the warming trend has been about 0.2°C per decade or 0.02°C per year, while year-to-year variations are about 0.1 to 0.2°C per year – five to ten times larger. Consequently, over periods of just a few years, the large year-to-year variations can dominate the longer-term trend. There have been many times since the nineteenth century that an extreme year was not surpassed for another decade or so.

Over longer periods, the year-to-year variations average out to zero, while global warming continues to push temperatures up. Over periods of several decades, the global warming signal dominates over internal variability. This is why scientists always analyze time series that are several decades long in order to assess the impact of climate change. One must certainly not draw conclusions from the few years immediately after an extremely hot year like 1998. In fact, Figure 3.1 shows that as warming has continued, the high temperature that was extraordinary in 1998 is rapidly becoming commonplace, and in all likelihood will soon be left behind by even more extreme years.

Note that all these variants of the claim the Earth is not warming (when they are not simply false) are based on selecting a small part of the available data – a single data source, a limited region, or a few years following an exceptional

[4] There are small differences between the two standard records of global temperature trends. The GISS (NASA) record shows 2005 as slightly warmer than 1998 but within the error bars, while 2007 is tied with 1998; the Hadley Centre (UK) record shows 2005 and 2007 both slightly cooler than 1998, both also within the error bars.

year – and claiming that this refutes the evidence in the rest of the record. It is this selective use of data that makes all these claims misleading or wrong. The conclusion that the Earth is warming rests on an overwhelming volume of multiple data sources, over the entire Earth over an extended period, which cannot be refuted by any such limited selection from the data.

Claim 2: the Earth may be warming, but the cause is climate variability, increased intensity of sunlight, or some other natural process – not human activities.

This argument accepts the evidence of warming, but denies its attribution to human causes. Like the denial that warming is happening, this claim comes in a few related variants. One variant accepts that warming is caused by atmospheric CO_2, but says the increase in atmospheric CO_2 comes from some natural source, not fossil fuel use. This claim goes against all available evidence. Emissions from fossil fuels can be observed going into the atmosphere, and the atmospheric increase matches the fossil-fuel source in quantity, isotopic mix, and timing. For the atmospheric CO_2 increase to be some natural process, it must be a process whose timing and quantity has neatly tracked human fossil-fuel use, and must come with an explanation of where, other than the atmosphere, all the carbon from burning fossil fuels has gone. This makes no sense at all.

Other variants of this claim accept the increase of atmospheric CO_2 as man-made, but reject this increase as the cause of observed climate change. This requires an alternative natural cause for the observed warming, of which two are most frequently proposed: natural variability, and increased solar activity. Natural variability advocates have argued that twentieth-century warming is a continued recovery from the last glacial maximum 20,000 years ago, or a recovery from the more recent several-century cool period called the "little ice age," or part of a naturally occurring 1,500-year cycle.

As Section 3.2 discussed, natural variability has been rejected as a significant contributor to recent warming for several reasons, including the fact that neither the record of pre-nineteenth century climate change, nor climate models without greenhouse-gas emissions, show any patterns of variability that resemble the rapid recent warming. The specific claim that recent warming is a "recovery" from some prior cool period presumes the climate has some normal state it returns to after warmer or colder periods, like a stretched spring returning to its normal length. This might seem like common sense, but it has no foundation in either the record of past climate or the physics of the atmosphere. There is no natural restoring force that pushes the climate warmer after a cool period like the little ice age. All climate change, natural or man-made, must have an underlying physical cause to account for it, but those advancing this argument usually do not identify a mechanism. And in fact, the only mechanism that has been identified that can account for recent warming is the effect of elevated greenhouse gases. As Section 3.2 discussed, all known natural mechanisms have been thoroughly investigated and have failed to account for the observed recent warming.

When a non-human mechanism is proposed, it is usually that the Sun has become brighter. There are many reasons to reject this claim, but two stand out. First, solar output has been measured since the 1970s, and it shows no trend since then that could account for more than a tiny fraction of the observed warming. Second, solar-driven climate change would warm both the lower atmosphere and the stratosphere, but the observed trends are warming in the surface and lower atmosphere and cooling in the stratosphere. In other words, what we are seeing is consistent with warming from greenhouse gases, but not with warming from the Sun.

Moreover, any claim that current warming is a natural phenomenon must explain why the obvious source of the warming, greenhouse-gas increase, is not causing warming. Recall that the evidence for greenhouse-gas warming stands on three strong foundations: basic physics says greenhouse gases should cause warming; warming is occurring; and models of greenhouse-gas effects quantitatively reproduce the observed warming. In view of this evidence, any claim that some other factor is causing the warming has to meet two basic requirements. It must show how the proposed mechanism is operating; and it must also explain why greenhouse gases, with their known absorption properties, are not causing the warming expected from them. No claimed alternative cause has succeeded at either part of this requirement. Other natural factors have obviously driven large climate variations in the past, and may be playing some role in recent warming – but that role is at most a small one.

> *Claim 3: future warming will be small. Even if human activities have caused recent warming, further warming this century and beyond will lie near the bottom of the projected range, or even below it.*

Like the previous two claims, this one also comes in two variants. The first asserts that climate sensitivity is much lower than now believed, so the Earth will not warm much even if atmospheric abundances of greenhouse gases keep growing. The second asserts that current emissions projections are too high: emissions will grow slowly or even decline, even with no efforts to limit them.

As discussed in Chapter 1, climate sensitivity likely falls in the range 2.0 to 4.5°C. Of this, the direct radiative effect of greenhouse gases accounts for 1°C, while the remainder, and the uncertainty, come from feedbacks in the climate system. The most powerful of these is the positive feedback from water vapor, which roughly doubles the direct effect of greenhouse gases: warming increases evaporation, which causes further warming because water vapor is a greenhouse gas. The current sensitivity range lies well above 1°C because the balance of evidence and scientific judgments indicates that the net effect of all climate feedbacks is positive and strong. For sensitivity to lie far below 2°C, there would have to be a large negative feedback somewhere in the climate system that has not been identified.

What might this be? The most plausible candidates involve changes in the properties of clouds. Different cloud types at different altitudes exert different climatic effects, some contributing net warming and others net cooling. In total, clouds presently cool the Earth by 20 to 30 W/m^2. If cloud cooling were somehow to increase sharply as the climate warmed, this could offset much of the warming from increased greenhouse gases. Over the past ten years, several proposals for such a large negative feedback have been made, but little supporting evidence has been found. In addition, it is hard to reconcile the existence of a large negative feedback, which would strongly stabilize the climate, with the large warmings and coolings that are seen in the climate record over the last 100 million years.

The argument is not fully closed, but the bulk of evidence continues to indicate that the net feedback is strongly positive. Advocates of strong negative feedbacks have sometimes suggested the argument is settled in their favor and current warming projections are consequently much too high. These claims greatly misrepresent the state of scientific knowledge. The question is not fully settled, so it remains possible that some large negative cloud-feedback will be found, but the balance of evidence and expert judgment lie much closer to the opposite conclusion, that there is no such feedback. The stated sensitivity range of 2.0 to 4.5°C, which is quite wide, is a fair representation of the current distribution of expert opinion on this uncertainty.

Alternatively, some writers have claimed climate will not change much because emissions will grow little, if at all, even without efforts to control them. This is not exactly a scientific claim, but one of substantial significance for the urgency of limiting emissions. The best evidence to support this claim is that through several years in the 1990s, global emissions grew slower than had been projected by prior scenarios. But since 2000, emissions growth has tracked at or above the highest emissions scenarios from the late 1990s, and recent shifts back toward higher-carbon fuels suggests these scenarios are more likely to be too low than too high.

Nevertheless, scenarios of future emissions span a wide range precisely because there is so much uncertainty about the factors driving them, as we discuss in the next chapter. Moreover, since it is well established that people tend to estimate uncertain quantities too confidently, the true uncertainty in future emissions may well be even wider than current scenarios represent. While it would be wonderful if emissions in fact declined without active intervention, there is no basis for confidence in this and a responsible approach to the climate issue must consider a wide range of potential emissions futures. Moreover, stabilizing climate change requires not just that emissions stop growing, but that they decline greatly from present levels. Even the lowest emission scenarios in Figure 3.14 do not stabilize the climate. So even if this claim turned out to be correct, it would not mean that climate stabilization could be achieved without active reduction efforts; it would only make them easier.

It is not possible to address all the erroneous and misleading claims advanced in the climate debate, so we have concentrated on summarizing a few prominent and recurring ones. In addition, these claims are a moving target. Their proponents have typically retreated step-by-step as advancing knowledge has moved their claims from being merely unsupported to ridiculous. For example, while political commentators and editorials still occasionally claim the Earth has not warmed over the past century, most scientific deniers have retreated from this claim over the past few years – later than the evidence warranted, to be sure, but still an indication of their need to maintain some degree of scientific credibility. When a few years hotter than 1998 have been recorded – which in all likelihood will happen soon – the claim that global warming stopped in 1998 will probably also disappear, no doubt to be replaced by a new one.

Finally, we should note that there are ample opportunities to use biased, misleading, and erroneous scientific arguments on all sides of policy debates. In preparing this book, we looked hard for prominent, purportedly scientific claims from environmental activists as biased or misleading as those summarized here from deniers, but found few. Some advocates have made claims of definite connections between single weather events, such as a particular hurricane or heat wave, and climate change; given the variability of climate and weather, such connections can generally not be made reliably. A few individual activists have made insupportably strong claims about severe human-health impacts from climate change, including statements that current climate change is already implicated in recent resurgence of infectious diseases. In fact, future climate change could bring large health impacts, but the uncertainties are large and the evidence for a strong climate signal in current health trends is ambiguous and contested. A few other advocates have implausibly exaggerated the technological options already available that would allow mitigation at zero or negative cost – although technological progress could well turn these present wild exaggerations into future realities. Finally, a few advocates and numerous press accounts have described potential catastrophic impacts in terms that imply these are near-certainties, rather than significant but uncertain risks.

But this does not amount to much. Climate-change statements of the major environmental organizations are quite careful, and there is nothing on the environmentalist side resembling the cottage industry of climate-change deniers and supporting organizations who publicly dispute the reality of climate change and its attribution to human activities. It is certainly possible to exaggerate environmental risks relative to scientific knowledge, and environmental advocates have sometimes done it. But in the present climate-change debate, the weight of misrepresentation appears to lie strongly with the deniers of climate change and the policy actors who use their arguments to oppose greenhouse-gas mitigation.

3.6 Conclusions

We conclude by summarizing the answers that present scientific knowledge provides to the four key questions about global climate change.

Is the climate changing? Yes, it is getting warmer. Multiple independent data sources confirm beyond any doubt that the Earth is warming, and that today's temperatures are rising particularly rapidly. In fact, we have high confidence that global-average surface temperatures over the past few decades are higher than any comparable period over the last five centuries, and possibly over the last thirteen centuries.

Are human activities responsible? It is very likely that greenhouse-gas emissions from human activities have caused most of the rapid warming of the past few decades. For warming prior to the middle of the twentieth century, emissions from human activities probably played a role, but natural processes such as solar variability, volcanoes, and internal climate variability could have also made substantial contributions.

What further changes are likely? Under all plausible scenarios for the twenty-first century, it is virtually certain that the climate will continue to warm, and the best-guess estimate of global warming this century ranges from 1.8 to 4.0°C. Even the bottom end of this range is more than double the warming of the twentieth century. Including the entire range of uncertainties widens the range of projected changes this century to 1.1 to 6.4°C.

What will the impacts be? We have a broad idea of the types of regional changes and impacts that are likely, but cannot predict specific impacts with confidence. The range of possible future impacts includes some that are serious enough to compel our attention. If climate change lies near the low end of the projected range (Fig. 3.14), impacts over the twenty-first century are likely to be manageable for rich, mid-latitude countries, but may pose serious difficulties for poorer countries. If climate change lies near the top of the projected range, impacts this century are likely to be severe and potentially unmanageable for everyone. Impacts of continued unabated climate changes after 2100, with atmospheric CO_2 increasing beyond triple or quadruple the pre-industrial level, are even more uncertain, but include a non-negligible risk of catastrophic changes that would fundamentally transform human ecosystems and societies.

Further reading for Chapter 3

ACIA (2004). *Impacts of a Warming Arctic: Arctic Climate Impact Assessment.* Cambridge, UK: Cambridge University Press.

This report is a synthesis of the key findings of the Arctic Climate Impact Assessment (ACIA) and is written in plain language accessible to policy-makers and the general public. The ACIA is a comprehensively researched and peer-reviewed assessment of Arctic climate change and its impacts for the region and for the world. It was written by an international team of hundreds of scientists, and also includes the special knowledge of indigenous people. This synthesis, as well as the full report, are available online at http://www.acia.uaf.edu.

K. Emanuel (2007). *What We Know About Climate Change.* Cambridge, MA: MIT Press.

This relatively short book outlines the basic science of global warming and how the current understanding has emerged.

IPCC (2007a). *Climate Change 2007: The Physical Science Basis.* Contribution of Working Group I to the Fourth Assessment Report of the Intergovernmental Panel on Climate Change, S. Solomon, D. Qin, M. Manning, Z. Chen, M. Marquis, K.B. Averyt, M. Tignor, and H.L. Miller (eds.). Cambridge, United Kingdom and New York, NY, USA: Cambridge University Press, 996 pp.

This is the most recent full-scale report of the IPCC's Working Group I, the group responsible for international assessments of the atmospheric science of climate change. It is the most recent authoritative statement of the status of scientific knowledge about climate change, and an essential source for anyone wishing to be literate in the climate change debate. In addition to the fully detailed and cited syntheses of specific aspects of climate-change science presented in each chapter, the report includes a technical summary and a policy-makers' summary that present the most important results and conclusions in more condensed and accessible form. We draw extensively on this report for many of the scientific conclusions we present in this chapter.

IPCC (2007b). *Climate Change 2007: Impacts, Adaptation and Vulnerability.* Contribution of Working Group II to the Fourth Assessment Report of the Intergovernmental Panel on Climate Change, M.L. Parry, O.F. Canziani, J.P. Palutikof, P.J. van der Linden, and C.E. Hanson (eds.) Cambridge, UK: Cambridge University Press, 976 pp.

This is the most recent full assessment of the IPCC's Working Group II, which summarizes present knowledge about potential impacts of climate change, ability to adapt, and vulnerability of environmental and social systems to climate change.

IPCC (2007d). *Climate Change 2007: Synthesis Report*. Contribution of Working Groups I, II and III to the Fourth Assessment Report of the Intergovernmental Panel on Climate Change, Core Writing Team, R.K. Pachauri, and A. Reisinger (eds.). Cambridge and New York: Cambridge University Press.

This report summarizes and integrates the principal results of the three IPCC working groups into a single volume.

US Climate Change Science Program (2006). *Temperature Trends in the Lower Atmosphere: Steps for Understanding and Reconciling Differences*, T.R. Karl, S.J. Hassol, C.D. Miller, and W.L. Murray (eds.).

This peer-reviewed assessment describes the numerous details of a calculation of the global average temperature. The trends in several important data sets are compared, and it is concluded that the trends are all generally consistent, although some discrepancies do exist.

US Climate Change Science Program (2009). *Global Climate Change Impacts in the United States. Unified Synthesis Product*, T.R. Karl, J. M. Melillo, and T. C. Peterson (eds.).

This report, produced by the US Climate Change Science Program, provides a detailed assessment of potential climate-change impacts, vulnerabilities, and capacity for adaptation for the United States. Separate studies examine effects of climate change on nine major US regions and seven sectors of national importance, updating the last comprehensive assessment of US impacts of climate change and variability published in 2001. Like the IPCC reports, this assessment involved the work of hundreds of scientists and was subjected to a rigorous and thoroughly documented process of peer review.

S. R. Weart (2003). *The Discovery of Global Warming*. Cambridge, MA: Harvard University Press.

A highly readable and accessible history of major developments in the science of climate change, from the nineteenth century through the formation of the modern consensus about the reality and predominant human cause of recent climate change as expressed in the 2001 IPCC report.

4

Climate change policy: impacts, assessments, and responses

Understanding the science of climate change provides only part of what is needed to decide what to do about the issue. Deciding how to proceed also requires information about the likely impacts of climate change on human society, the options available to respond to climate change, and the tradeoffs these options present in terms of effectiveness, benefits, costs, and risks. This chapter summarizes present knowledge and uncertainties on these matters.

The responses available to deal with climate change fall into two broad categories, called *adaptation* and *mitigation*, plus a third type of potential response, only recently receiving serious attention, called *geoengineering*. Adaptation measures target the impacts of climate change: they seek to adjust human society to the changing climate, to reduce the resultant harms. Examples include building sea walls or dikes to limit risks from higher sea levels or river flooding, or planting drought-resistant crops to deal with drier summers in agricultural regions. Mitigation measures target the causes of climate change: they seek to slow or stop climate change by reducing the emissions of greenhouse gases that are responsible.

In early climate debates, many advocates treated mitigation versus adaptation as an either-or choice, perhaps because they imagined attention and support for the response they favored would be weakened by acknowledging the need for the other. Fortunately, debate has moved past this false dichotomy, and it is now widely understood that both adaptation and mitigation are needed. Adaptation is necessary because climate change is already occurring and substantial further change is unavoidable. We will have to adapt to these impacts, no matter how aggressively we limit emissions. Mitigation is necessary to limit the severity of climate change we must adapt to, because the most extreme projected climate changes, and therefore the most severe impacts, remain avoidable, particularly in the latter half of the century. The key questions in shaping a response are not which of these to do, but how much of each, how soon, and how to do both most effectively and efficiently.

Most proposed responses to climate change focus on mitigation and adaptation. The third type of response, geoengineering, involves actively manipulating the climate system to offset the effects of greenhouse gases, thereby breaking the link between emissions and climate change. Although geoengineering responses to climate change were proposed by a few scientists as early as the 1960s, they have received less attention than mitigation and adaptation since climate change came onto policy agendas in the 1980s. Geoengineering has seen a resurgence of interest in the past few years, however, as recognition has spread of the seriousness of climate change and the ineffectiveness of responses thus far. Despite this new interest, understanding of the benefits, costs, and risks of geoengineering remains limited and preliminary.

The chapter is organized as follows. Section 4.1 discusses impacts of climate change and adaptation measures. Section 4.2 discusses projections of emissions over the next century and the technologies and policies, nationally and internationally, available to reduce them. Section 4.3 discusses current estimates of the costs of climate impacts, mitigation, and adaptation, and efforts to integrate these into a consistent framework to evaluate responses. Section 4.4 briefly discusses geoengineering measures, while Section 4.5 summarizes current knowledge and expert judgments regarding beneficial responses to climate change, emphasizing how to make sensible climate-change decisions under uncertainty.

4.1 Impacts and adaptation

4.1.1 Defining and assessing the impacts of climate change

Chapter 3 discussed present knowledge of how the climate is likely to change over this century, and the resultant impacts on ecosystems and natural resources. But determining the impact of these changes on people requires additional analysis that links projections of climate change and its direct impacts on ecosystems and resources with an understanding of how society depends on these aspects of the climate and environment. As in Chapter 3, we are concerned in this section with positive questions of how to describe and analyze the interaction between climate-driven changes in physical or biological processes and human societies. Normative questions, concerned with how to value these changes once they are described and analyzed, and what they imply for decisions, are addressed in Section 4.3 (valuation) and Chapter 5 (implications for action).

The impacts of climate change on people or communities depend not just on how the climate changes, but also on multiple socio-economic factors related to where and how people live, how rich or poor they are, how they earn their livings, what technologies and natural resources they rely on, and what institutions, cultural practices, and policies govern them. Consequently, impacts will vary among people and places, not just because they are experiencing different climate changes but also because they differ in their sensitivity to specified changes. Particular people, places, and activities are sensitive to particular aspects of climate in particular ways. An electric utility might be sensitive to the length and frequency of summer heat waves, which raise electrical demand, while a ski resort might be highly sensitive to changes in average winter temperature and total snowfall but not at all to summer temperature. Farming in a particular country might be sensitive to changes in total growing-season precipitation and the frequency of heavy downpours and droughts. Low-lying coasts, whether in Louisiana or Bangladesh, are especially sensitive to sea level rise, while rapid-growth regions already facing water shortages, such as California and Arizona, are sensitive to changes in precipitation and winter snowpack. Consequently, projecting climate-change impacts requires linking projections of climate change with scenarios of the socio-economic factors that most strongly shape these sensitivities.

Impacts of climate change also depend on how narrowly or broadly you look. Look at smaller regions or narrower parts of the economy, and impacts are normally larger and more diverse, including both gains and losses. Even in one town, warmer and drier summers might harm agriculture but benefit tourism. If you step back to consider the town's whole economy, farming losses and tourism benefits partly cancel, making the total impact smaller. The more broadly you look, the more small-scale effects average out. National or global impact estimates thus combine and conceal variation in smaller-scale impacts, with some people, places, and activities being harmed more, while others are harmed less or even – for small climate changes – benefited.

Impacts of climate change also depend on what else changes at the same time. As the climate changes, other environmental factors will change in parallel: atmospheric CO_2 will certainly increase, and other factors such as nutrient deposition, air quality, and land-cover are also likely to change. Many human and biological systems are sensitive both to climate change and to these other changes, and to interactions between them. The most studied interaction is that between climate and CO_2. In addition to changing the climate, elevated CO_2 affects plants directly by increasing the efficiency of photosynthesis and water use, although the effects vary widely among plant species. Studies of agricultural crops have mostly found these effects of elevated CO_2 can offset the water stress of projected warmer, drier climates, so plant growth on balance increases under the small changes projected for the next few decades – although this increased growth appears to come with lower protein content and possibly smaller kernel size in cereal crops. The larger climate changes projected later in the century shift the balance between the climate and CO_2 effects to a net reduction in plant growth. Studies that go beyond average climate to include changes in variability and extreme weather, such as projected increases in both droughts and heavy downpours, show negative effects that may more than offset near-term increases from changes in CO_2 and average climate. Moreover, changes in climate and CO_2 will affect weeds, pests, and diseases as well as crops. Experimental studies of these interactions suggest that weeds and invasive species may respond more strongly to elevated CO_2 than crops, so total crop yields could go either up or down for particular species relationships (of which there are a huge number to be studied).

The biggest uncertainty in projecting climate impacts, however, is estimating how people will adapt to the changes. Societies are adapted to present climates in diverse ways, and we expect some adaptation to future changes. If climate change reduces yields and profits from current farming practices, we expect farmers – and others who influence their decisions, such as seed suppliers, equipment companies, and farm extension services – to shift to crops and practices better suited to the new conditions. If present settlement patterns, economic activities, or resource-management practices become ill suited to a changed climate, we expect people to notice this and, eventually and to some degree, change practices to better fit the new climate. Moreover, people do not have to wait for a change to occur to adapt to it. With good forecasts, people may look ahead and adapt in advance, either to specific changes they expect or to a general increase in climate uncertainty. Such anticipatory adaptation is especially important for decisions with long-term consequences, such as planning, zoning, and infrastructure policies or long-lived investments such as ports, dams, and power plants.

How people will adapt is crucial in determining climate impacts and must be considered in assessing them. Unfortunately, knowledge of how people adapt and what factors influence their capacity to do so is quite limited. Many impact assessments have relied on one of two extreme assumptions about adaptation. Some have assumed today's practices continue with no response to changed climate. By assuming zero adaptation, this approach systematically overstates harms from climate change. At the opposite extreme, some studies assume optimal adaptation, unconstrained by any limits of foresight or rational decision-making, or even by rigidities due to long-lived capital equipment. Just as assuming no adaptation predictably overstates harms from climate change, assuming perfect adaptation understates them. Indeed, by assuming society adapts better to future climate than today's society is adapted to the present climate, it is possible to project, implausibly, that the impacts of nearly any climate change will on balance be beneficial.

Better impact projections require realistic projections of how societies will actually adapt to future climate changes, but these are hard to do. The capacity for adaptation varies strongly among people and places. Rich societies with well functioning institutions and strong social networks are generally more adaptable, and so less vulnerable to climate change, than societies without these advantages. Having the capacity to adapt, however, does not necessarily mean using it to adapt well. There is ample evidence that adaptation to present climate conditions is far from ideal, even in rich and well-governed societies. We operate intensive agriculture in drought-prone regions through unsustainable mining of groundwater. We build in high-risk locations on low-lying coastlines, flood plains, and fire- and slide-prone hillsides. We even rebuild repeatedly in the same high-risk locations, often with large subsidies, after property is destroyed. Such maladaptations make us more vulnerable than necessary both to present climate variability and to projected future climate change.

As climate change continues and its impacts become more obvious, it may be that societies will pay more attention and adapt better, but this is not at all assured. Adaptation is highly context specific: effective adaptation measures depend on multiple environmental, social, economic, and institutional conditions in each setting, and many factors can complicate and obstruct adaptation. For example, decisions that affect vulnerability and adaptation are often not recognized as such. The relevant decisions may be motivated by multiple preferences and values, with vulnerability to weather or climate not the most prominent. People do not build on vulnerable hillsides for the purpose of putting themselves at risk, but because they want to be there for other reasons such as cost, convenience, amenity, or culture. Moreover, uncertainty in local and regional climate projections can make it hard for even well informed and motivated managers to know what changes they must adapt to, and so decide what adaptation measures to incorporate in their investment or planning decisions. These factors make effective adaptation difficult, and also obstruct accurate projections of impacts and vulnerabilities that depend upon adaptation.

In sum, projections of societal impacts of climate change are more difficult and uncertain than the projections of climate change discussed in Chapter 3. But no matter how uncertain impact projections may be, they still must be considered in any reasonable judgment of how to respond to climate change. The alternatives – ignoring impacts, or delaying action until we know what the impacts are – would effectively mean waiting until the changes are actually upon us, when our options would be much more limited.

Because acting under uncertainty cannot be avoided, there is great value in developing better impact-assessment methods that consider uncertainty, integrating presently available research with expert judgment where necessary. When such integration has been attempted, it has typically involved rolling available knowledge into some aggregate expert judgment. For example, authors of IPCC assessments in both 2001 and 2007 summarized their judgments of the likely severity of impacts in charts. The 2007 chart, reproduced as Figure 4.1, suggests that severe impacts are likely to mount with average global warming of more than 2.5 to 3°C beyond the level of 1990.

While such summaries of expert judgment are a useful start for summarizing climate-impact risks, providing more useful guidance for decision-making requires methods of synthesizing current knowledge and expert judgments that are more transparent in their underlying reasoning, better at incorporating uncertainties, and if possible more quantitative. The need for impact assessments that better integrate uncertainty is especially acute for possible abrupt changes. As Chapter 3 discussed, a few mechanisms of abrupt change have been identified as possibilities, including loss of major ice sheets in Greenland or Antarctica and large-scale rearrangement of ocean circulation. There may be other possibilities not yet identified, just as no one anticipated the Antarctic ozone hole until it was observed.

As larger climate changes are considered – whether due to higher emissions, higher climate sensitivity, or looking further into a future without emissions controls – the potential for abrupt changes and severe impacts appears to grow more likely, but how likely? Most experts have judged these abrupt changes to be unlikely this century, but opinion is mixed and there is no basis for confidently dismissing these as too unlikely to consider. Indeed, the past few years have seen substantial increase in expert concern about risks of major ice-sheet loss.

In fact, many observers judge that it is the low-probability risk of severe impacts, not the smaller impacts of more likely middle projections, that provides the main reason to limit climate change. A decision to exclude ice-sheet loss from sea-level rise scenarios in the 2007 IPCC Working Group I, made because the authors could not agree how to describe its likelihood, was one of the points of sharpest controversy in that assessment. But while ignoring such risks in assessments clearly seems wrong, including them poses the risk that they will dominate the assessment and resultant decisions, despite their believed low probability. Assessing such risks responsibly requires considering judgments of both their severity and their probability, but we lack widely accepted methods and processes for doing so, and for integrating these assessments into

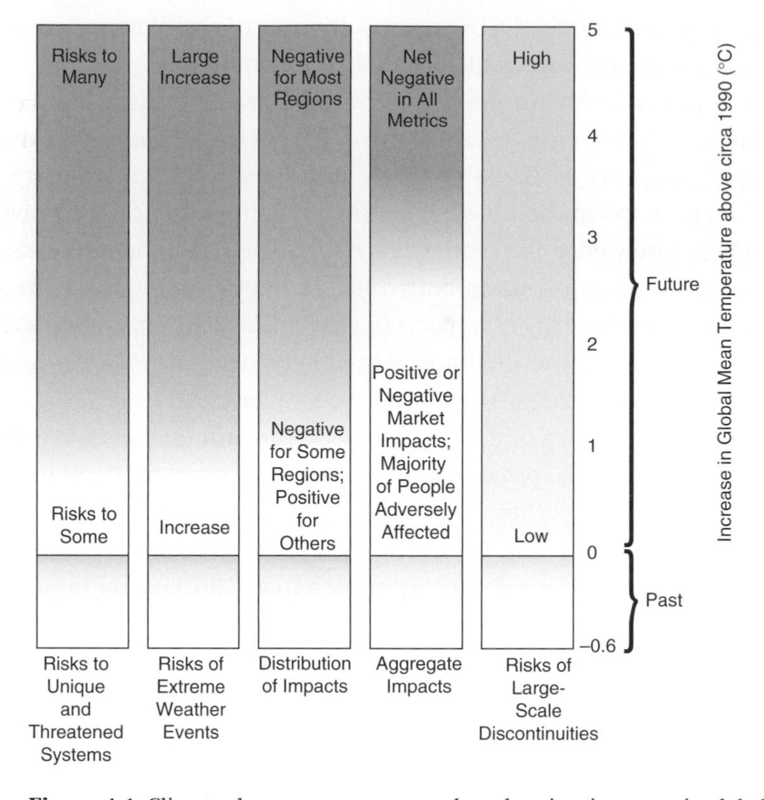

Figure 4.1 Climate change consequences plotted against increases in global mean temperature (°C) after 1990. Each column corresponds to a specific "reason for concern," and represents additional outcomes associated with increasing global mean temperature. The shading scheme represents progressively increasing levels of risk. The historical period 1900 to 2000 warmed by ~0.7 °C and led to some impacts. It should be noted that this figure addresses only how risks change as global mean temperature increases, not how risks might change at different rates of warming. Furthermore, it does not address when impacts might be realized, nor does it account for the effects of different development pathways on vulnerability. Source: adapted from Smith et al. [2009].

responsible decision-making. Further development and acceptance of such methods is a high priority.

4.1.2 Responses to enhance adaptation

Adaptation to climate change is not just the job of government, perhaps not even primarily the job of government. Much adaptation will be undertaken by individuals, businesses and other organizations, and communities, acting on their own behalf. But public policies and governments can help promote societal adaptation to climate change, in three ways. First, governments can provide information and assistance, such as climate projections, impact studies, assessments of potential responses, and technical and financial aid in implementing responses. Such support can help citizens and communities shift from reacting to changes as they occur toward anticipating future changes, so adaptation will be more effective and less costly, particularly when it can be integrated into planning and investment decisions with long time horizons. Better assessments and projections can also reduce the risk of wasting money and effort adapting to the wrong changes, e.g., adapting to some short-term climate fluctuation because it is mistaken for a long-term trend. Second, governments can use their regulatory authority to require citizens to reduce their vulnerability, for example by changing zoning codes to restrict building in vulnerable areas. Third, governments can use their own spending and operations to build adaptive capacity directly, e.g., by building dikes or other coastal defenses, or by requiring that infrastructure projects to which they contribute (e.g., roads and other transport, water and sewer systems, electric power and communications grids, ports) are designed with reduced vulnerability.

Because the specific changes to which we must adapt will remain uncertain, one job of adaptation measures is to increase society's robustness to a broader range of climate conditions. Many adaptation measures will not be specific to climate, but will simultaneously reduce vulnerability to multiple risks. For example, strengthening public-health systems will reduce health risks from climate change along with other health risks. Similarly, strengthening emergency-response systems and implementing policies to promote development and reduce poverty would reduce vulnerability both to climate change and to many other threats, including more immediate ones.

Measures to promote adaptation are an essential part of the response to climate change, because we have missed the chance to keep climate changes small. In addition to the warming that has occurred, we are already committed by past emissions to a further 0.4°C warming above 2008 temperatures, which would occur even in the impossible event of human emissions dropping to zero today. Moreover, even extreme efforts to reduce emissions are unlikely to hold warming below 1.5 to 2°C above today's temperatures. This range of further change, which is now essentially unavoidable, will carry impacts – serious ones for some people and some places – for which the only possible response is some combination of adaptation and simply enduring the resultant harms.

Effective adaptation will require money, including substantial public expenditures and international transfers. It will also require novel institutional structures, including new networks sharing knowledge, experience, and methods among local, national, and international institutions worldwide. And even with vigorous pursuit of adaptation, it is likely that many things people value – resources, assets, and especially ecosystems – will be lost. So one of the main jobs of adaptation will be buying time to assess and prioritize what can be saved, and what must be sacrificed.

Adaptation measures alone, however, cannot make up an effective response to climate change. Future trends in emissions will determine how much and how fast we drive climate change, and the more the climate changes the more severe the resultant impacts will be and the more adaptation capacity will be challenged. Relying on adaptation while doing nothing to slow or stop climate change would mean putting no limits on how much change we must adapt to – gambling that we can effectively, and at acceptable cost, endure or adapt to any amount of climate change. The evidence that this would be an imprudent gamble – including maladaptations to present climate, wide variation among people and societies in adaptation capacity, and the non-negligible possibility of extreme changes and impacts, particularly under high-emissions futures – is powerful. To limit the impacts we and our descendants must suffer or adapt to, it is also necessary to reduce the emissions that are causing human-driven climate change. The next section discusses this approach to managing climate change, and what is presently known about the technical and policy options available to pursue it.

4.2 Emissions and mitigation responses

4.2.1 Emission trends and projections

Section 3.3 discussed trends and scenarios of greenhouse-gas emissions from human activities used to project how the climate will change over this century. Here, we provide more detail about emissions trends and projections, the factors shaping them, and the means available to reduce them.

The largest source of the human emissions contributing to climate change is carbon dioxide (CO_2) released by burning fossil fuels – coal, oil, and natural gas. Human society relies heavily on these fuels, which provide about 80 percent of total energy use worldwide. In 2006, emissions of CO_2 from burning fossil fuels worldwide were about 8.0 billion metric tons of carbon (GtC).[1] Land-use

[1] A metric ton is 1000 kilograms or 2200 pounds, about 10 percent larger than a short ton. A billion metric tons (or a Gigaton) of carbon is often abbreviated 1 GtC. We state quantities of CO_2 as the mass of the carbon contained in them. An alternative convention, used by many sources, reports the total mass of the CO_2 molecule. Because the molecular weight of CO_2 is 44, while the weight of the one carbon atom in it is 12, 1 GtC of emissions is the same as 3.67 $GtCO_2$.

change, mainly deforestation due to land clearing, added roughly another 1.5 GtC (although this figure is substantially more uncertain), bringing total anthropogenic emissions of CO_2 to about 9.5 GtC – a global average of about 1.4 tons per person, given a world population of 6.8 billion.

Several other gases emitted from various industrial and agricultural activities also contribute to climate change. The most important of these are methane (CH_4), which is emitted from rice paddies, landfills, livestock, and the extraction and processing of fossil fuels, as well as several natural sources; nitrous oxide (N_2O), which is emitted from nitrogen-based fertilizer and industrial processes as well as several natural sources; and the halocarbons, a group of synthetic industrial chemicals used as refrigerants and in various other industrial applications. Although emitted in much smaller quantities than CO_2, these other greenhouse gases contribute more warming per ton emitted and so add significantly to the total warming effect of CO_2. These gases also differ from CO_2 in their atmospheric lifetimes, which determine for how long current emissions contribute to warming. For example, methane emissions remain in the atmosphere for only about ten years on average, while the most stable halocarbons remain for many thousands of years. Emissions of the major greenhouse gases have been growing since the industrial revolution, with the largest increases over the past few decades. Of the total warming effect from increased greenhouse gases over the past two centuries, CO_2 contributes about two thirds, while these non-CO_2 greenhouse gases make up the other third.

Still other human pollutants and activities are also changing the radiation balance of the atmosphere and influencing climate change, in ways that are more complex in their mechanisms and more uncertain in their total magnitude. Changes in atmospheric ozone, which human activities are increasing in the troposphere (lower atmosphere) but decreasing in the stratosphere (the layer above the troposphere, starting about 10–15 kilometers up), on balance make a warming contribution about 20 percent of that of increased CO_2. Various human pollutants and activities are raising the atmospheric abundance of aerosols, small solid or liquid particles suspended in the atmosphere that have short atmospheric lifetimes and contribute a mix of warming and cooling effects. One important human source of aerosols is sulfur dioxide (SO_2) – a pollutant from burning sulfur-containing fuels, especially coal – which forms fine liquid droplets that reflect incoming sunlight and cool the surface. Another important aerosol is black carbon, sooty material from incomplete combustion that absorbs sunlight and warms the surface. Aerosols are presently making a net cooling contribution about equal to the warming from non-CO_2 greenhouse gases, but the uncertainty in this estimate is large. Overall, the human activities and pollutants contributing to climate change are diverse, but CO_2 from fossil-fuel combustion represents more than half of the total.

There are large differences among nations in how much they are contributing to climate change, now and over time. The industrialized countries, with about one-sixth of world population, are responsible for about 55 percent of today's world CO_2 emissions. Long the largest source of emissions, the United States currently emits about 20 percent of world CO_2, but rapid economic growth in China brought its emissions to surpass those of the United States in 2007. The industrialized countries' share of world emissions is larger if you consider cumulative historical emissions (about 73 percent of total fossil-fuel CO_2 emissions since 1950), and smaller if you consider not just fossil-fuel CO_2 but also land-use change and other greenhouse gases (about 44 percent of 2000 emissions).

Section 3.3 introduced the six standard scenarios of potential twenty-first-century emissions trends produced by the IPCC in the late 1990s. These six scenarios show world CO_2 emissions, now about 9.5 GtC per year, ranging from less than 5 GtC to more than 30 GtC in 2100. This large range indicates substantial uncertainty about emissions trends and resultant climate change, although subsequent analyses have suggested that emissions near the bottom of this range would be quite unlikely without intentional efforts to reduce them. Moreover, no explicit interpretation of the uncertainty represented by these scenarios was provided. It is not stated, for example, if the authors judged emissions 90 percent likely to lie within this range, as opposed to, say, 99 percent likely or even certain, nor even whether they judged the middle of this range to be more likely than the extremes.

Despite these wide uncertainties in future emissions trends, most recent scenarios show a few strong regularities. First, baseline scenarios do not lead to climate stabilization: indeed, most baselines show continued emissions growth through this century, as global economic growth outpaces emission-reducing technological innovations. Recent analyses typically project global emissions in 2100 around 20 to 25 GtC, about triple current emissions. Second, all scenarios show developing-country emissions overtaking those of the presently industrialized countries within the next few decades. Third, scenarios show that emissions are highly sensitive to trends in energy resources and technologies. Plausible variation in these factors – particularly alternative assumptions of where world energy moves as cheap conventional oil and gas decline, toward coal and high-carbon synthetic fuels or toward non-emitting energy sources – produce as wide a range of emissions futures as any plausible assumptions for future population and economic growth.

Although scenarios are mainly produced to support longer-term assessment and planning, divergence between scenarios and observed emissions trends has

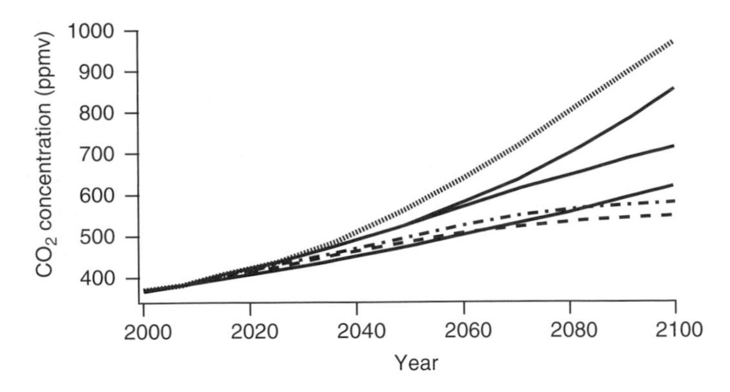

Figure 4.2 Projected abundance of CO_2 in the atmosphere based on the IPCC emission scenarios described in Section 3.3, in parts per million. The dotted line is scenario A1FI, the dashed line is scenario B1, the dot-dashed line is scenario A1T. *Source*: adapted from Figure 18 of the Technical Summary, IPCC [2001a].

generated controversy over the past two decades, as emissions in the 1990s grew slower than prior scenarios had projected, then grew faster than even the highest-growth scenarios since 2000. Recent high growth has raised concern that scenarios may be understating the emissions pressure from the shift now underway toward higher-carbon fuel sources such as coal, heavy oils, and oil sands, which is projected to continue as conventional sources are depleted and prices rise.

Figure 4.2 shows the trends in atmospheric CO_2 that follow from the six IPCC marker scenarios. Trends from all scenarios are similar for the next few decades, passing through 500 parts per million (ppm) around mid-century, but diverge later in the century. By 2100, the lowest scenarios have CO_2 around 550 ppm (double the pre-industrial value) and growing very slowly, while the highest scenarios have CO_2 around 900 ppm (triple the pre-industrial value) and still rising steeply. These large differences in atmospheric CO_2 will bring large differences in climate change, but these will also only grow large late this century and thereafter.

Like CO_2, other greenhouse gases are projected to increase through the century in most scenarios, and to shift toward developing countries. Projections for SO_2 and other aerosols are more mixed. Overall, most scenarios project declines in SO_2 over the century, as they are controlled to limit acid rain and other regional pollution. Since these emissions now contribute a regionally concentrated net cooling, reducing them will make an additional contribution to climate warming. Projected trends differ strongly over time and by region, however, with large near-term increases projected in rapidly industrializing regions for a few decades, followed by reductions thereafter.

In view of the complexity of projecting multiple types of emissions and aerosols, scenarios are increasingly being described in terms of total human contribution to climate forcing, rather than counting each type of emission separately. Total climate forcing is the change in total heating of the surface by solar radiation and by infrared radiation from the atmosphere. As Chapter 1 discussed, adding greenhouse gases to the atmosphere increases the heating of the surface by the atmosphere. The net climate forcing by human activities (greenhouse gases, aerosols, plus other smaller effects) is presently about 1.6 watts per square meter – 2.6 watts from CO_2 and other long-lived greenhouse gases, partly offset by a negative forcing of about 1 watt from aerosols and other effects. Uncontrolled emissions scenarios recently produced for the IPCC and the US Climate Change Science Program (CCSP) project that human forcing from increased greenhouse gases will increase to 6.5 to 8.5 watts per square meter by 2100, equivalent to a CO_2 concentration of 920 to 1390 ppm.

These scenarios all represent informed guesses of how world emissions might grow under plausible, consistent assumptions about trends in population growth, economic growth, and technological change – and excluding any disruptions or surprises. The scenarios were developed to provide emission inputs for model projections of how the climate is likely to change. But emission scenarios can also be constructed in a different way, to serve a different purpose. They can instead be built around environmental targets, and used as tools to examine the targets' feasibility and evaluate alternative ways of achieving them. The targets in such scenarios are usually expressed as some limit on atmospheric change, such as a limit on global-average temperature, radiative forcing, or concentration of greenhouse gases.

Of these possibilities, a temperature limit is most closely linked to changes in climate and impacts, but because climate sensitivity is uncertain, the limit on greenhouse gases needed to meet a temperature limit is uncertain. Similarly, if greenhouse gases or radiative forcing are held to a fixed limit, the resultant temperature change will be uncertain. For example, one recent analysis found that holding greenhouse gases to 450 ppm CO_2-equivalent – about the strictest limit now proposed – would give only about a 50 percent chance of limiting warming to 2°C above pre-industrial temperatures. Due to this uncertainty, scenarios have usually expressed stabilization targets in terms of concentration or radiative forcing. Early examples considered alternative stabilization levels for CO_2 alone, typically 450, 550, 650, and 750 ppm. (Recall that CO_2 has increased from about 270 to 380 ppm over the past 200 years, and is now increasing about 2 ppm per year.) As discussions have moved to jointly limiting all greenhouse gases, stabilization scenarios have increasingly been defined in terms of radiative forcing. For example, recent US CCSP scenarios considered four stabilization levels ranging from 3.4 to 6.7 watts above pre-industrial (530 to 970 ppm CO_2-equivalent), while new scenarios for the next IPCC assessment are examining

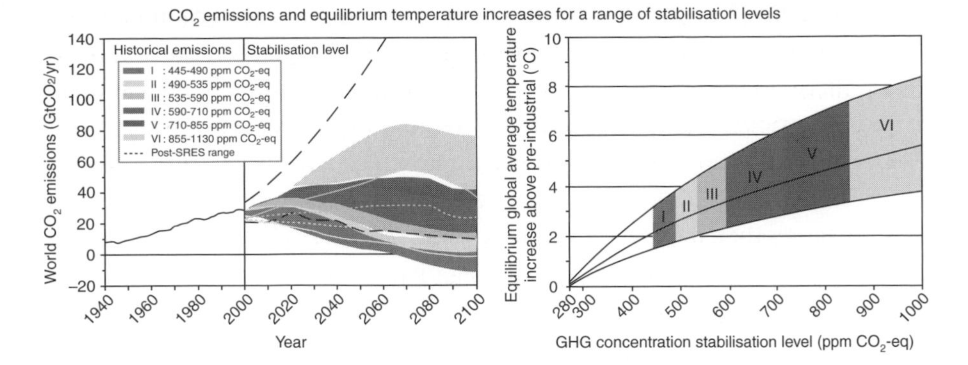

Figure 4.3 The relationship between emissions trajectories and climate stabilization. The right panel shows the range of global temperature change associated with different levels of atmospheric greenhouse-gas stabilization. The left panel shows ranges of emissions scenarios associated with meeting various atmospheric-concentration stabilization targets. The lowest band of scenarios shows that stabilizing around 2°C warming requires holding concentrations to 450–500 ppm CO_2-equivalent, which in turn requires that global emissions decline 50 to 85 percent by mid-century.
Source: Figure SPM.11, IPCC [2007d].

stabilization at 6.0, 4.5, and 2.6–3.0 watts (equivalent to 850, 650, and 450–490 ppm CO_2-equivalent).[2]

Stabilization scenarios tend to have similar trajectories of global emissions over time, initially rising, reaching a peak, and declining. The tighter the stabilization target, the sooner the peak occurs and the more sharply emissions fall thereafter. Figure 4.3, for example, summarizes many stabilization scenarios compiled by the IPCC. In scenarios that aim to stabilize at 450–500 ppm CO_2-equivalent, emissions peak before 2015 and decline by 50 to 85 percent (from 2000 levels) by mid-century. Relaxing the stabilization target to 500–550 ppm delays the emission peak by a decade or so and reduces the required mid-century cuts to 30–60 percent. Further relaxation of the target allows still later peaks and slower reductions thereafter.

Stabilization emission paths do not have to have this shape. There are many paths to each target, including some that start cutting immediately and others that make larger cuts starting later. But this shape, with gradual deflection of near-term emission growth followed by larger reductions, tends to reduce the cost of stabilization for several reasons. It avoids premature scrapping of long-lived capital equipment such as power stations; it allows more time to develop

[2] Two candidates are still being considered for the strictest control scenario to be used. Both peak at 3.0 watts during the century, then decline to 2.6 to 2.9 watts by 2100.

new low-emitting technologies; and by delaying emission-reduction expenditures, it reduces their present value through discounting.[3] As targets become stricter, there is less flexibility in how to meet them. Strict targets may require starting reductions immediately, and may require rapid broadening of controls to all types and sources of emissions, in particular emissions from land-use change. The strictest targets may require letting concentrations initially overshoot the target then decline, through extreme reductions in emissions or even measures to drive net human emissions negative, such as recapturing CO_2 from the atmosphere.

Whether an emissions scenario is constructed as a projection or a goal, a picture of emissions alone can provide only a starting point for thinking about mitigation strategy. Understanding why emissions might follow one path or another in a projection scenario, or identifying what efforts are required for emissions to track a target scenario, requires asking what factors cause emissions to change one way rather than another, and what measures are available to deflect their trends. The next section turns to these questions.

4.2.2 Factors underlying emission trends

A first step to understanding causes of emission trends and ways to reduce them is to decompose emission trends into trends in the underlying factors discussed in Chapter 3: population, economic growth (GDP per person), and technology (CO_2 emitted per dollar of GDP, which can be further decomposed into energy per dollar and CO_2 per unit energy). These factors influence each other, of course, so this accounting exercise of decomposing the factors does not mean each factor can be varied independently, or even that any factor can necessarily be controlled: all the factors are aggregate descriptions of socio-economic trends. Over the past few decades, world population growth has declined to slightly over 1 percent per year due to sharp reductions in fertility rates through much (not all) of the world, while world GDP per person has grown 1 to 2 percent per year on average. Technological change has tended to

[3] Discounting is the process of converting costs or benefits that occur at different times to a common scale so they can be aggregated and compared. Because of the productivity of capital resources – dollars, trees, or fish today can make more dollars, trees, or fish next year – future quantities are normally *discounted*, treated as equivalent to smaller quantities today. Mechanically, discounting normally involves multiplying by a constant factor per time period, equivalent to a constant compound interest rate on a savings account. Discount rates used to evaluate public policies and investment projects typically range from about 1 percent to 10 percent per year. Rates used to evaluate proposed investments in the private sector are usually higher. The role of discounting in evaluating climate-change responses is discussed in Section 4.3.2.

reduce emissions, in that CO_2 emitted per dollar GDP declined a little more than 1 percent per year on average over the twentieth century. The net effect of these trends was that CO_2 emissions grew about 1 percent per year on average over the twentieth century.

Any strategy to slow or reverse emission growth must achieve some combination of shifts in the trends of these underlying factors: a faster decline in world population growth; slower economic growth; or an acceleration of technological innovation. But policies explicitly aimed at limiting population are contentious in the extreme. All emission scenarios discussed above assume continued growth in world population, but at a declining rate. But because the causes of recent fertility decreases are weakly understood, there is uncertainty over how effective policies to accelerate the trend could be, even disregarding the deep political, religious, and cultural controversies such policies provoke. If fertility declines slow or reverse, emission growth could easily exceed even the top of the present scenario range.

Explicit policies to limit economic growth for the sake of the environment are, if possible, an even more explosive topic than policies to control population growth. Relieving the extreme poverty of many of the world's citizens provides a compelling rationale for worldwide growth in economic output, provided the income growth actually reaches those who need it. But even in rich societies, a central focus of policy remains the promotion of economic growth, and there is no evidence that people's desire to consume more exhibits any satiation, with the single exception of food. Both within and between countries, economic growth also fulfills a sharper political need: the prospect of continued growth can mute political pressure to resolve inequities and social problems, by giving the disadvantaged hope that their lot will improve if they wait. Although it has long been suggested that aggregate material consumption must eventually stop growing – by writers from classical economists to modern ecological economists and development convergence theorists – this argument has gained little traction in policy debates. As a result, mitigation policies always focus mainly on promoting technological change, to reduce emissions per dollar of economic output.

Historically, the world economy has shown a clear long-term trend of decreasing CO_2 emissions per dollar of GDP. This decline has occurred because of improvements in both the energy intensity of economic output (energy per GDP, or how much energy it takes to produce one dollar of economic output) and the carbon intensity of energy (CO_2 per unit energy, or how much CO_2 is emitted to deliver one Joule of energy). Energy per GDP has declined due to both increased efficiency in particular production processes and a shift toward less energy-intensive activities (less steel and other heavy industry, more secondary manufacturing and services). Carbon intensity of energy has also declined, through a gradual shift from higher-emitting energy sources (wood, then coal) toward lower-emitting ones (petroleum and natural gas, with some movement to nuclear and renewables). In the past decade, however, this decarbonization trend has stopped worldwide and reversed in some regions, as peak production of conventional oil has been approached (or by some estimates, passed) and energy supply has begun moving toward heavy and unconventional sources such as oil sands, and back to coal.

Further reductions in either how much energy the economy consumes, or how much CO_2 is emitted to deliver this energy, could greatly reduce future emissions, as the huge gap between technologically optimistic and pessimistic emissions scenarios (even with the same population and economic growth) shows. But what rates of improvement are plausible? As the next section discusses, further progress in decarbonizing the energy system is possible through expanded use of natural gas to replace oil and coal and shifting new energy supply to non-emitting sources. But current energy-market conditions are not supporting this shift, as the recent reversal of the long-term trend shows. Further reductions in carbon intensity are technically achievable, but will take concerted efforts.

Energy intensity of GDP has at times decreased as fast as 2 percent per year, but the periods of fastest decreases have reflected special conditions, such as periods of rapid shift in economic mix or responses to energy price shocks. Consequently there is basis to doubt that such high rates can be sustained for decades or extended to the world. For the whole world over the twentieth century, the average decrease was more modest, about 1 percent per year. Prospects for future progress look similar in many areas of energy conversion and end-use. Large efficiency improvements have been made, whether through accumulation of many small gains (e.g., improvements in electric motors and internal-combustion engines) or through discrete, high-impact innovations (e.g., high-efficiency compact fluorescent lighting and low-emissivity windows). Large further gains are possible, including some from innovations already identified (e.g., LED lighting) or even immediately available and cost-effective, and others that are more distant, difficult, and uncertain.

Most analysts judge that 1 percent average annual improvement can be sustained through the century, while some suggest that rising energy prices or other incentives could raise the rate of sustained annual improvement to about 2 percent. Aggregate gains from efficiency improvements may be limited, however, by linkages with economic growth. Because efficiency gains are usually realized in investments in new capital equipment, high rates of improvement require strong economic growth. Moreover, because efficiency gains can make energy services (e.g., driving, heating, or lighting) cheaper, people may respond to the gains by consuming more. Estimates of this "rebound effect" vary widely, from 10 to 20 percent offset of the initial energy savings in some studies to more than 100 percent in others.

Disaggregating the factors underlying emissions trends helps clarify the size of the challenge of stabilizing climate. The emissions cuts required for stabilization – 30 to 60 percent reductions by 2050 to stabilize at 500–550 ppm, or 50 to 85 percent cuts to stabilize at 450–500 ppm – imply sustained annual emissions reductions of 1 percent for the weaker goal to 1.5 percent for the stronger. Assuming a requirement of, say, 4 percent annual economic growth over this period, which might be made up of 0.5 to 1 percent population growth and 3 to 3.5 percent per capita income growth, implies the need for 5 to 5.5 percent annual decline in the two technology coefficients together, i.e., 2.5 percent annual decrease in *both* energy per GDP and carbon per unit energy.

4.2.3 Technological options to reduce emissions

Comparing these required rates of change to historical experience shows what a huge challenge emission reductions of this magnitude will be. But there is a large and diverse array of known technological options available to pursue them, plus many further prospects more remote in time, cost, or confidence of success.

While energy-related CO_2 is the largest share of human greenhouse-gas emissions and thus provides the largest opportunities for cuts, other emissions – including CO_2 from agriculture, forests, and land-use, and non-CO_2 gases including methane, nitrous oxide, and halocarbons – also provide significant reduction opportunities. These opportunities are diverse: some emission sources such as methane from landfills, feedlots, and natural gas pipelines are concentrated, easy to identify, and readily reducible, while others are decentralized, hard to monitor, and mainly reducible through changes in management or behavior such as agricultural practices, rather than through technology. These sources offer substantial opportunities to broaden emissions cuts and so reduce the cost of achieving any specified target. For strict stabilization targets this cost advantage grows large, to the point that the strictest targets, such as 450 ppm CO_2-equivalent, may not be feasible unless land-use emissions and non-CO_2 gases are included. In many cases, however, the means to control these emissions and the policies to motivate these controls are harder to identify and enforce than for energy-related CO_2 emissions.

For energy-related CO_2 emissions, the most immediate reduction opportunities lie in increasing efficiency of energy use, thereby reducing energy use per GDP. Engineering analyses typically find many such opportunities available at low and even negative cost, through initiatives such as more efficient building, heating, lighting, and electronic equipment, and conservation of hot water. One recent analysis found opportunities to cut US emissions 20 percent at negative marginal cost and more than 40 percent at marginal cost below $180/tC.[4] There is long-standing evidence of such seemingly attractive efficiency options not being adopted, however, suggesting either hidden costs not captured in the engineering analyses, or market failures obstructing adoption. Large gains may also be feasible in transportation, from further efficiency improvements in current drive-train technologies and from switching to new energy carriers such as electricity or hydrogen, although the cost and the actual gains achieved would depend on the whole energy system – i.e., how are the electricity or hydrogen produced – not just the vehicles.

Additional opportunities, particularly in the longer term, come from reducing carbon intensity, emissions per unit energy delivered, through shifting energy supply toward low or non-emitting sources and technologies. The major types of climate-safe energy technology are well known, and include renewable sources such as solar, wind, and biomass; nuclear fission and perhaps in the future nuclear fusion; and carbon sequestration, by which CO_2 from fossil-fuel combustion is captured and stored in biological or geological reservoirs rather than released to the atmosphere.

Renewable energy sources already provide several percent of world energy. The two biggest renewable sources – firewood and hydroelectricity – cannot be expanded much more, however, while the remaining sources – solar, wind, geothermal, and ocean (thermal, tides, and waves) – together contribute only about 1 percent of present world energy. Wind and solar power are expanding rapidly worldwide, but from a very small base. They are already cost-competitive in some small niche applications, principally remote locations off the electricity grid, while modern large-scale wind turbines in sizes up to several Megawatts are increasingly competitive even in centralized power systems on windy sites, allowing wind to supply 10 to 20 percent of electricity in a few countries – although the countries where wind and solar are expanding most rapidly give them large subsidies.

Continuing incremental innovations to increase conversion efficiencies and reduce costs can allow further expansion of these sources. There are also

[4] The marginal cost of a specified reduction in emissions is the cost of cutting the last ton to reach that specified total reduction. In calculus terms, the marginal cost is the derivative, or rate of change, of the total cost relative to the quantity reduced. The marginal cost of a specified reduction normally differs from the average cost, because the first few units of reduction are the cheapest to achieve, while additional cuts grow more difficult and costly. Consequently, as emission cuts are increased, marginal costs increase faster than average costs, which continue to include the effect of the early cheap cuts.

prospects for expanded use of biomass energy, cultivating fast-growing plantation crops and efficiently burning them to provide energy with no net CO_2 emissions, provided the plantations are managed sustainably. Current biofuel technologies are highly variable in their sustainability, however – some rely heavily on external fossil-fuel inputs – and large-scale expansion of biofuels would risk competing for land with food production and preservation of forests and biodiversity.

Most renewable sources, however, suffer from two limitations that obstruct their ability to expand to a large fraction of world energy use. They have low power densities, so providing a lot of power requires installations covering large areas, both to collect the resource and to transport the energy to demand centers. For example, to meet a substantial fraction of US electricity demand with solar power would require a solar array of some thousands of square kilometers, most likely in a desert location far from major load centers. Biomass is limited by photosynthesis to even lower power density, about 0.6 W/m^2, so providing a substantial fraction of world energy this way would require about as much land cultivating energy crops as is now used for agriculture worldwide. Moreover, solar and wind provide energy only intermittently – when the Sun is shining or the wind is blowing – so they need backup or energy-storage systems to deliver energy reliably all day and all year. While a huge expansion of renewable energy is feasible in principle – estimates of maximum practical supply range as high as 30 TW, double today's world energy supply – such expansion would depend on solving these problems of siting, transmission, and energy storage.

The renewable source that would best avoid the problems of low power density and intermittency would be solar energy collected on arrays in space and transmitted to receiving stations on Earth. Because sunlight is stronger and always available in space, the required array area would be only about one-tenth that needed on the Earth's surface to deliver the same power. Acting against this advantage is the high cost of launching material into space, presently many thousands of dollars per kilogram. Because this is a large fraction of the total estimated cost of space solar systems, projected large reductions in launch costs could sharply reduce total system costs – making them competitive with solar power on the surface, or even with fossil fuels – but these projected reductions remain speculative.

Like renewable sources, nuclear fission and fusion are energy sources that emit no CO_2 to the atmosphere. Nuclear fission reactors, which generate energy by splitting uranium or plutonium atoms, have been in large-scale use worldwide for decades. Construction of new reactors stalled in the 1970s, however, due to concerns over safety, waste disposal, terrorism, and the risks of nuclear weapons proliferation from diversion of reactor fuel. Fusion reactors, which generate energy by fusing two hydrogen atoms to create a helium atom, remain in development after decades of research.

Nuclear power could make a large contribution to world energy by mid-century, subject to major remaining obstacles and uncertainties. For fission, new reactor designs hold the promise of greatly improving safety, while the waste-disposal problem appears likely to be solvable technically, if perhaps not politically. The most acute challenge to large-scale expansion of fission remains the security risks – sabotage, terrorism, and diversion of fuel to weapons fabrication – which may not be surmountable. It has also been suggested that world uranium resources may not be adequate to sustain a large-scale fission industry without chemical reprocessing of fuel, a process likely to further increase the risk of illicit diversion of fuel to make weapons. Fusion remains a speculative resource, still awaiting the technical breakthroughs that must precede commercial viability, so no significant contribution from it can reasonably be expected for at least several decades.

The final major technological route to reducing CO_2 emissions is to burn fossil fuels, but in a way that releases little or no CO_2 to the atmosphere, through technologies for carbon capture and storage (CCS). There are several promising approaches. One of these involves decomposing the fossil fuel before it is burned into its major chemical constituents, hydrogen and carbon. The hydrogen is burned to provide energy, emitting mostly harmless water vapor. The carbon is buried in a long-term reservoir underground or undersea. Recent progress in these technologies suggests that this approach is technically viable, is compatible with present energy systems, and would cost substantially less than present renewable or nuclear sources. The key uncertainty about this approach is the availability and long-term stability of the CO_2-storage sites. If the storage is not stable, so stored carbon returns to the atmosphere too fast – faster than a few thousand years, on average – then the approach will be ineffective. Early research suggests that some sequestration sites, including depleted oil and gas fields, deep salty aquifers, deep coal seams, and perhaps the deep ocean for certain chemical forms of carbon, are reliably stable for much longer periods. Although the safety and stability of these reservoirs needs further research and careful assessment of associated risks, CCS presently appears to have substantial promise to reduce emissions, particularly over the next few decades when fossil fuels remain the primary source of world energy.

Carbon can also be sequestered biologically, in trees or soils, although the size and longevity of these reservoirs, and their vulnerability to sudden events like forest fires or rapid decomposition associated with global warming make them look less promising than geological sequestration. Systems that combine growing biomass for energy production with separating and sequestering the resultant carbon, as well as systems that recapture CO_2 directly from the atmosphere, also show promise.

Overall, the prospects for sharply cutting world emissions through technological innovation are mixed. On the one hand, cutting emissions to stabilize climate change is a technically solvable problem: there are multiple feasible technological routes to provide energy without greenhouse-gas emissions. On the other hand, none of these sources is free of problems or potential conflicts. A vast expansion of climate-safe energy sources is needed to stabilize climate at reasonably prudent levels. Under mid-range scenarios of energy demand, the required new supply of climate-safe energy by 2050 may range from roughly equal to today's total world energy supply, to two or three times as much.

At such scales of deployment, all new sources will face tight supply bottlenecks and even the most seemingly benign sources will bring substantial environmental impacts and associated political opposition. Moreover, the mitigation problem is so big we cannot limit our response to the technologies we now prefer: we cannot tell what mix of technologies will turn out to be successful and socially acceptable, so in view of the potential problems with all, excluding any major candidate in advance – in particular, excluding nuclear or CCS from the mix – would risk our ability to achieve strict climate-stabilization goals. Finally, except for the low and negative-cost opportunities in conservation and efficiency improvement – which probably only capture a small part of the problem – climate-safe energy sources cost more than the conventional, emitting sources they would replace. Consequently, they will not be deployed – certainly not at the rapid rate required – without policies to motivate or require them. The next section discusses what form these policies might take.

4.2.4 National policy responses

Decisions to develop and deploy climate-safe energy technologies will mostly not be made by governments for the purpose of slowing climate change. Rather, like most economic decisions, they will be made by thousands or millions of individuals and organizations for their own diverse purposes, responding to their own perceptions of their present opportunities, costs, and risks, and their guesses about future ones. Government policy plays a key role in influencing these millions of private choices, however, by enhancing private actors' capacity to make socially preferred choices, providing information to facilitate such choices, and crucially, by changing their perceptions of the opportunities, costs, and risks that motivate their choices – i.e., their incentives.

Many types of public policy can influence these decisions and so change emissions trends, including broad tax and fiscal policy, but four types of policy are particularly relevant and explicitly targeted at emissions reductions. These include market-based regulatory instruments such as emission taxes or tradable emission permits, conventional regulations, direct public expenditures, and various initiatives that rely on information, education, and voluntary actions.

Market-based mechanisms

Market-based regulatory mechanisms are the most prominent new environmental policies adopted over the past 10 to 20 years. These policies pursue environmental goals by providing incentives operating through markets, to which individuals have flexibility in choosing their response. To control emissions of greenhouse gases or other pollutants, there are two main forms of market-based policies: an emission fee or tax, generally called a "carbon tax" when applied to greenhouse-gas emissions, or a tradable emission-permit system, often called a "cap-and trade" system. Under a carbon tax, each source must pay a specified charge for each ton of pollution emitted. Under a cap-and-trade system, each source must hold a permit for each ton it emits. The government distributes permits initially, after which emitters may buy and sell them among themselves.

The advantage of these market-based policies is the flexibility they grant to emitters in how to respond. They do not specify how much any particular source must cut, but let each one choose how much to emit, as long as they either pay the tax or hold permits for the emissions they choose. The central effect of either policy is to make emissions costly. Each emitter faces a cost for each ton they emit, which should motivate them to reduce their emissions and avoid the cost. Under a tax, every emitter will cut until the marginal cost of the next ton is equal to the tax rate: up to that point, they would rather make the cheaper cuts available to them than pay the tax, while beyond that point they would rather pay the tax than make the more expensive cuts available to them. Under cap-and-trade, the market price of permits will be set by trades in which emitters with higher marginal costs, who would rather pay for a permit than make the more costly reductions available to them, buy permits from those with lower marginal costs, who prefer to make the cheap cuts available to them to receive the price from selling a permit. If the policy is set at the right level of stringency – the tax rate or the number of permits distributed – and if emitters respond rationally to these incentives, then the socially optimal configuration of emissions will result, in terms of both overall emissions and the distribution of emissions among sources.

These two forms of policy are often proposed as the central element of a climate-change mitigation strategy. A carbon tax would be charged on fossil fuels in proportion to their carbon content. The tax could be levied at various points in the energy system, from initial fuel extraction to the point of emission. One approach, often proposed for ease of administration, is to apply the tax "upstream" where the fuel is imported or produced (i.e., at the coal mine or oil well), with a tax credit generated whenever fuel is diverted into a non-emissive use such as petrochemical manufacture or long-term sequestration. The tax would then be embedded in the price of the fuel as it passes through the economy, raising the cost of all goods and services that use fossil energy. A cap-and-trade system would be implemented in a similar way: a permit for the carbon content would be required to extract or import a unit of fossil fuel, while a new permit would be generated for each unit of carbon stably sequestered or incorporated in a non-emissive use. The cost of the permit, like the carbon tax, would follow the fuel through the economy, raising the price of carbon-based goods and services.

Alternatively, a carbon tax or cap-and-trade system can be implemented "downstream," at the point where the fuel is burned and the CO_2 emitted. This is the approach of the EU emission trading system, and the cap-and-trade proposals now being developed at the state and federal level in the US. To be administratively feasible, downstream systems have to limit their scope to a relatively small number of large, stationary emissions sources, usually electrical generating stations and large industrial facilities. This has the effect of narrowing the share of the economy that faces the resultant emissions price.

Emission taxes and cap-and-trade systems both aim to provide consistent incentives for mitigation, but they differ from each other in a few important ways. Because an emission tax is charged on every unit of emissions, taxes transfer wealth from emitters to the government. A permit system would make the same wealth transfer if, as often proposed, emitters must buy permits in an auction. In existing systems, however, permits have usually not been auctioned, but given for free to current emitters. Implemented in this way, cap-and-trade systems are much less costly to current emitters and so meet less opposition and are easier to enact.

The effects of emission-tax and cap-and-trade systems also differ when there is uncertainty about the costs and benefits of cutting emissions. A permit system fixes the total quantity emitted, regardless of how much it costs to reduce to that level. A tax system fixes the cost of the last unit of emission to be cut – because emitters will cut until it is cheaper to pay the tax than cut further, then stop – regardless of how much emissions are actually reduced to reach this point. Consequently, when the costs and benefits of cutting emissions are uncertain, which system is preferred depends on which of these quantities – the total quantity reduced, or the marginal cost of reductions – is more important to get right.

In current debate over these two approaches, there is a growing disconnect between the judgments of economists and other analysts, and the main direction of political action. Experts are increasingly identifying carbon-tax approaches as preferable to cap-and-trade systems, for several reasons. First, because climate change depends on cumulative emissions over years to decades rather than emissions in any particular year, a carbon tax can be adjusted over time as needed to direct emissions toward a specified climate-stabilization goal, while still putting a clear limit on mitigation costs in any year. Second, a carbon tax allows the intensity of the emission-cutting incentive to be precisely specified and varied over time. In contrast, cap-and-trade systems have experienced high short-term price volatility, which has focused attention on short-term trading opportunities and obscured incentives for the long-term innovations and investments needed to reduce emissions. Third, cap-and-trade systems are more likely to create valuable assets for private actors, and so pose greater risks of transfers, conflicts, and corruption, particularly in an international system with varying standards of implementation and enforcement. Yet despite these seemingly strong advantages to tax-based systems, virtually all current political proposals for market-based mitigation policies are cap-and-trade systems.

One way to gain some of the advantages of both carbon-tax and cap-and-trade systems is to construct hybrid or blended systems. Several forms of these are possible. A cap-and-trade system can include price limits to bound price volatility and private rents – either a price floor at which the government buys back permits if the cap is set too loose, or a price ceiling or "safety valve" at which the government sells additional permits if the cap is set too tight and the price unexpectedly high. Other hybrid systems might include an emission tax with quantity bounds, so the tax is raised if emissions or their growth rate exceed some threshold, or lowered if emissions fall faster than expected; or a tax charged not on all emissions, but only on emissions above some baseline, with rebates for emissions below the baseline.

Whether market-based policies take the form of a carbon tax, a cap-and-trade system, or a hybrid, the policy determines both an aggregate level of emissions reduction, and an associated price on emissions or marginal cost. All analyses of climate-stabilization scenarios have shown that the marginal cost, the price on emissions, *must* rise over time to provide the incentives for long-term investment and technology innovation needed to bring emissions down, although different models' estimates of the required emissions price vary substantially, especially after mid-century.

Conventional regulations

Before the recent interest in market-based policies, the most common environmental policies were regulations specifying some performance target that each emitter – e.g., a factory, a firm, or a product – must meet. Performance targets can be defined in various ways, e.g., total emissions of a pollutant per year, concentration of a pollutant in emissions, or emissions per unit of operations (e.g., regulations for automobile exhaust are defined in grams of each pollutant emitted per mile driven). A few environmental regulations have specified not just performance targets but also particular technologies or processes to achieve them, but this is less common. Conventional regulation of greenhouse gases would involve limiting the emissions of specific plants or types of equipment.

Regulations of this type brought large environmental improvements over the past 30 years, but were criticized for costing more than needed to achieve a specified environmental benefit. There are two causes of this inefficiency. First, when uniform targets are imposed on a group of emission sources (e.g., each plant must cut by 20 percent), emitters may differ in their marginal cost of reductions. When this occurs, it is possible to gain the same environmental benefit more cheaply by shifting reductions among sources, cutting more where cuts are cheaper (marginal control costs are lower), and less where they are more costly (marginal costs are higher). Performance standards also give inadequate incentives for emission-reducing innovations, because much of the benefit of such innovations comes from lowering the cost of reductions in excess of the standards, which emitters under a performance target are not required to make. These criticisms have sometimes been overstated, but are substantially correct and are the reason that conventional performance standards have largely been eclipsed by market-based mechanisms since 1990.

Public expenditures

Government expenditures can be instruments of environmental policy in several ways, for example by buying green products (e.g., efficient vehicles) for government operations even when these carry a cost premium. The largest role for direct public expenditure in mitigation policy, however, is government-supported research and development (R&D) of advanced energy technologies. There are strong arguments for public investment in energy R&D – both as a way to facilitate emission reductions, and to correct the market failure that arises from the public-good character of R&D, whereby firms cannot capture the whole benefit of knowledge produced by their research and so invest too little in it. Government investment in climate-safe energy R&D is particularly necessary because the long time horizons and large risks involved deter private investments. Yet despite widespread recognition that technological innovation is the major route to managing climate change, spending on energy research has been declining for at least the past decade in most industrialized countries. Many studies over decades have recommended large increases in federal

R&D expenditures on improving energy efficiency, renewable energy, nuclear energy, and carbon capture and sequestration, but the first significant increase in many years came in the stimulus package of early 2009.

Information, education, and voluntary measures

A final category of public policy tools includes educational, information-based, and voluntary measures. These measures seek to influence emissions by educating citizens and emitters about climate change and ways to reduce emissions. They may seek to help guide beneficial choices, or to motivate, coordinate, or honor voluntary mitigation efforts by firms. For example, policies requiring that cars and appliances carry labels explaining their energy efficiency allow consumers to consider this factor in their buying decisions. Providing climate forecasts can also motivate people to consider the need to adapt to a changing climate. Policies of this type can sometimes deploy real incentives. For example, requiring public reporting of emissions can motivate firms to reduce them, as can voluntary programs that carry prizes, public recognition, or linkages to government purchasing. But there are limits to the effectiveness of these policies, particularly when they must stand alone. Voluntary and informational measures usually cannot motivate changes that carry substantial costs or require large investments.

These four types of policy provide the building blocks of a complete mitigation strategy. There are many different ways to design the details of each type of policy, and combine the policies into a viable mitigation strategy – one that effectively limits emissions, at limited cost and administrative burden, and is feasible and sustainable in the relevant political setting. Essentially all analyses of mitigation strategies have agreed that measures to put a price on emissions, as broadly and consistently as possible across the economy, are the essential and central element of a mitigation strategy. Similarly, there is virtual unanimity that the emissions price must rise over time, by increasing the tax or decreasing the permit cap. A clear, pre-announced schedule of increasing stringency gives investors a stable planning environment and assures them that investments in reducing emissions will pay off. There is also wide agreement that public support for climate-safe energy R&D is necessary to complement the price on emissions, because of the public-good character of research.

The role of the other two types of policy in a mitigation strategy is more uncertain and contested. If incentives carried through the economy in energy prices are sufficient to motivate optimal responses throughout the economy, then no additional regulatory measures are necessary: market-based mitigation policies alone can achieve optimal, cost-minimizing mitigation. But if these energy-market incentives are not effective enough – e.g., if the scope of market-based policies is limited to some set of large industrial emitters, or if energy-market incentives are ineffective in some sectors due to split incentives, information limits, or other market failures, or if some emission areas are dominated by government action or other regulation rather than market-based decisions – then additional regulatory measures may be warranted. Commonly proposed areas that might need such additional measures include building efficiency codes, vehicle and appliance efficiency standards, and infrastructure investments. Finally, information-based and voluntary policies have often been derided as ineffective measures, adopted for symbolic purposes when governments are not serious about mitigation. In most existing cases, such policies have made little difference. But these can make a contribution under some conditions, complementing the effect of other, stronger policies by helping private actors recognize, understand, and respond to the incentives embodied in other policies.

4.2.5 International policy responses

The discussion of policy options thus far has focused on the national level, because this is where the strongest direct regulatory authority over emitters lies. But because greenhouse-gas emissions anywhere contribute to climate change everywhere, and because no one country dominates world emissions – even the largest emitters, the United States and China, each contribute slightly less than one quarter of the total – mitigation efforts must be coordinated internationally to be effective. Such coordination is difficult because in the international arena, no one is in charge. There is no world government with the authority to enact, implement, and enforce policy, or to compel governments to participate. Rather, international policy is made by negotiation among national representatives, often with industry and environment groups and other non-state actors present and trying to influence the outcome. This process is weaker, more cumbersome, and slower than national policy-making, but it is all that is available to respond to global problems such as climate change. Negotiators seek agreement on national commitments and efforts to address climate change, including their form, level, and the associated distribution of burdens and costs among nations over time – who will do what, when. Effective management of the issue requires that these national commitments be sufficiently strong, widely supported, and well designed to limit global emissions at minimal cost; that the associated burdens be distributed in a manner acceptable to all participants; and that these measures be implemented, enforced, and reviewed and adapted appropriately over time in response to experience and new knowledge and capabilities.

National commitments

While national policies can impose obligations directly onto emitters, international policies can usually only impose obligations on national governments, and only with their consent. Like domestic policies, national commitments to international policies take several forms. Governments can commit to performance targets like limits on national emissions, with varying degrees of flexibility in how to achieve them. Alternatively, they can commit to enact national policies, such as emission taxes or other forms of regulation. Or they can submit to international processes to motivate and enable national actions, such as reporting and exchange of information, assessment of national policies, or reviews of their progress. These three approaches are similar to three types of domestic mitigation policy discussed above, conventional regulation, market-based mechanisms, and voluntary and information-based approaches. Governments also have the additional option of creating international institutions to which some authority over policies or their implementation can be delegated.

Thus far, national emission targets have been the most common form of international mitigation commitment, used in both the Framework Convention on Climate Change (FCCC) and Kyoto Protocol as well as several other major environmental treaties. National targets have the advantages of being simple, clear, and familiar. Moreover, by defining clear responsibilities but leaving the means of implementation up to national governments, targets let governments be held accountable for their commitments with minimal intrusion on their sovereignty.

But national targets also have serious disadvantages. Because most emissions come not from governments but from citizens and businesses, a national target has no concrete effect until implemented in domestic policy. But the effects of domestic policies are uncertain, so governments cannot know in advance how hard or costly it will be to meet an emissions target, or even whether it is possible. Even in domestic policy, decades of failure to attain the air-quality standards in the US Clean Air Act show the potential gulf between adopting a target and achieving it. This uncertainty about targets' attainability introduces a sharp tension into international target negotiations, between making targets challenging and creating strong incentives to meet them. If governments face only minor consequences for missing a target, they have little incentive for serious efforts to meet it. But if the consequences are severe, governments will likely only agree to targets they are highly confident of meeting – weak targets, or ones with wide loopholes. Moreover, while clear, demanding targets can be good motivators, they do this best when nations are near the boundary between meeting and not meeting them. Incentives to exceed a target you already expect to meet, or to narrow the gap when you are clearly falling short, are much weaker.

Making national targets more effective policy tools would require extending the range of outcomes over which they provide incentives, while keeping the penalty for shortfall small enough that governments are willing to accept ambitious targets. One approach to achieve this is to allow flexibility in how targets are achieved, providing some of the cost-saving and incentive-broadening benefits of market-based policy mechanisms. Three types of flexibility have been proposed for international greenhouse-gas mitigation, called "what," "when," and "where" flexibility. "What" flexibility lets nations distribute their mitigation efforts among greenhouse gases, by defining targets jointly for emissions of multiple gases with each one counted by its contribution to climate change. "When" flexibility lets nations distribute their efforts over time to meet a long-term mitigation goal, by defining targets as multi-year emissions budgets rather than year to year.

"Where" flexibility is the most potent and most controversial form of flexibility. It lets nations exchange emissions (or obligations to reduce them) so a nation with high marginal mitigation costs can cut less than its commitment, and instead pay another with lower costs to cut beyond its commitment. "Where" flexibility could be implemented as a cap-and-trade system in which emissions permits are bought and sold internationally rather than just domestically, either through transactions between governments or directly between emitters in multiple nations. Like domestic cap-and-trade systems, "where" flexibility allows nations to shift mitigation effort to where it is cheapest, thereby reducing total costs by as much as 80–90 percent in some calculations, in addition to the cost savings provided by "what" and "when" flexibility. International flexibility mechanisms pose serious challenges of policy design and implementation, however, such as how to connect the mitigation obligations of national governments and individual emitters to maintain consistent incentives and accountability, and how to ensure the integrity of a trading system operating across multiple jurisdictions with widely different standards of implementation and enforcement.

An alternative to explicit national emission targets is to negotiate national policies, such as emissions taxes or other regulatory measures. Negotiating over policies instead of targets avoids the problem of uncertain achievability, since the taxes or other policies being negotiated clearly lie within governments' authority to enact. As in domestic policy, emitters under a tax would cut until their marginal cost was equal to the tax. The level of the tax would consequently determine the marginal cost of mitigation, and an equal tax in different nations would yield equal marginal costs and a cost-minimizing distribution of mitigation effort – but the emission reductions achieved would be uncertain.

As in domestic policy, there are strong reasons to favor emissions-tax systems, but the political momentum lies with negotiation of national targets and cap-

and-trade systems. Emission taxes also face several difficulties that are specific to the international level. The complexity and diversity of national tax systems would require intensive negotiations over implementation to prevent national loopholes, such as exemptions or other special treatment granted to carbon-intensive export industries. An internationally negotiated carbon tax is also more likely to raise sovereignty-based objections than negotiations over emission targets. Hybrid systems combining elements of both emission taxes and tradable-permit systems, as have been proposed for domestic policy, could help manage the uncertainties associated with each system separately and so ease negotiation difficulties. Nations could also negotiate other forms of policy – e.g., performance targets or other forms of regulation applying to particular emissions sectors, selected either for large contribution to emissions or for perceived need for consistent policies internationally to avoid trade distortions.

A third proposed approach to international mitigation commitments is for national governments to negotiate procedures, institutions, and rules, rather than committing to either specific policies or specific targets. This approach was considered in the climate negotiations of the early 1990s, under the slogan "pledge and review." Under this proposal, governments would pledge to enact mitigation policies of their own choosing, and state what results they expected. They would then be subject to periodic international review of the design and implementation of their policies, and the results projected and achieved. This approach sought to counter the objection that governments would reject binding policy commitments as infringements on their sovereignty. Pledge-and-review aimed to create and exploit less formal and stringent, but still potentially effective, incentives for national policy-makers, such as wanting to appear competent and responsible, and avoiding international criticism and embarrassment. In negotiation of the Kyoto Protocol, pledge-and-review was viewed as not tough enough and abandoned in favor of binding national emission targets.

Like voluntary and information-based measures in domestic policy, such procedural commitments can effectively complement other forms of commitment. Indeed, these may be necessary for effective implementation of other commitments. But as in domestic policy, exclusive reliance on this approach would probably be ineffective because it does not create strong enough incentives for all the major actors, national governments and others, whose efforts are required.

A final approach to international policy is to create institutions with authority to enact climate policy directly. Often proposed by those most skeptical of the competency or resolve of national governments to address climate change, this approach draws on the historical example of the negotiated formation of real trans-national power in the European Union, and on the specter of some potential future environmental calamity that would demand strong, perhaps even dictatorial, leadership. But in its strongest form, the approach would face serious obstacles. Excluding the possibility of *coup d'état* or revolution, the only way to establish such authority would be by negotiation among national governments. But if governments cannot agree to enact specific climate-change policies, how would they agree to give away their authority on the issue completely to some international body? Moreover, even if such international authority could be created, it is not clear how it could be controlled to ensure competency and democratic accountability. Effective international institutions in specific issue areas can grow over time through successive negotiated decisions. The present international trade regime, organized around the World Trade Organization (WTO), evolved in this way. The WTO provides a plausible model for a climate regime that would combine agreement on targets, policies, procedures, and international institutions to which specific authorities are progressively delegated as experience is gained and nations' confidence in the process grows. But it is not practical to contemplate this approach as a single, large-scale decision that would resolve the present international deadlock on climate.

Timing, distribution, and sequencing of commitments

Choosing the form of national commitments is not enough to make international action. It must also be decided who does how much, and when. The global scale of the climate issue means that effective mitigation requires participation by all major emitting nations, eventually. An effective global regime cannot be constructed in one step, however, but must be phased in over time. It is widely agreed that minimizing disruption and costs requires commitments to start weak then grow more stringent over time, with ample notice and with the ability to adapt in response to new knowledge and changed conditions.

The distribution of efforts among nations raises more contentious issues, which cannot be avoided. The deepest controversy over participation has concerned sequencing of mitigation commitments by industrialized and developing nations: must all major nations, industrialized and developing, undertake simultaneous mitigation commitments from the outset, or should the industrialized countries go first?

The arguments for including developing countries in mitigation commitments from the start are based in considerations of cost control and effectiveness. Many low-cost mitigation opportunities appear now to lie in developing countries, particularly those experiencing rapid economic growth, so a cost-minimizing distribution of mitigation would locate substantial near-term efforts there. It has also been argued that if the rich countries act alone, this would be made ineffective by "emissions leakage," as emission-intensive industries relocate to countries not controlling emissions. Such movement of investment would weaken near-term mitigation, because some emissions would be moved instead of being reduced, and would also make it hard to subsequently expand the control zone, by raising the cost of subsequent mitigation in those initially non-participating countries to which high-emitting investment moved. So instead of any group moving ahead, all nations should move together from the outset, even if this means the action they take is very weak, then gradually move together to steeper cuts.

On the other hand, treaties can expand participation over time, sometimes very quickly. Concern about emission leakage rests on an assumption that the movement of high-emitting investment to non-participating nations occurs much faster than participation in the mitigation regime expands. The limited experience gained so far with building international environmental regimes suggests otherwise. The Montreal Protocol provides the most relevant example. Initially negotiated with stringent CFC controls for a relatively small group of nations – and widely criticized at the time for the risk that CFC-based industries would move abroad – the ozone treaty moved rapidly both to expand participation and to increase the stringency of controls. Details of the treaty's control measures, especially its restrictions on trade in relevant products with non-parties, served both to deter the outflow of investment and to motivate additional nations to join. Although the details will be harder, it is likely that greenhouse mitigation commitments could also be designed to minimize incentives for investment outflow and to promote expansion.

The arguments for the contrary approach, initial actions by the rich, industrialized countries, with the developing countries beginning mitigation efforts later, are based in part on provisions of the Framework Convention on Climate Change and arguments of fairness. Today's rich countries got rich by heavy use of cheap fossil fuels and so are responsible for most of the excess greenhouse gases now in the atmosphere. But these arguments are also based in considerations of political practicality. The current long-standing deadlock, with both industrialized and developing countries demanding that the other side move first, is only likely to be broken by a first, serious step by some core group of participants willing to agree to it. In all likelihood this must be the industrialized countries, but the costs and risks of such a first step need not be severe. With phased commitments in which policy stringency increases gradually, the costs of such a first step would necessarily start small, bringing very limited competitive burdens and risks of leakage.

And to the extent that any serious movement – including a program that starts small but includes real commitments to increasing stringency over time – is likely to help break the deadlock, others may be willing to undertake commitments soon thereafter, particularly if developing countries' commitments can be made conditional on signs of serious efforts from the industrialized countries. There may be ample room for negotiation here, particularly if parties recognize that where emission-reducing investments are located is a separate matter from who pays for them. In practical terms, there are likely to be many plausible agreements that couple mitigation with negotiations on technology, finance, and other issues that allow exchange of benefits so all participants gain. Of course, any approach that even starts with unequal marginal costs in different regions will be inefficient, but if the only way to achieve effective action is to endure some such period of sub-optimality, this is preferable to continued delay that means doing nothing. International environmental diplomacy is an area where the perfect can be, and in many debates thus far has been, the enemy of the good. Rather than seeking optimality in early steps, it should be guided by some simple principles of pragmatic benefit: early actions should make an effective contribution to the problem, should make subsequent expansion of efforts easier, and should not lock in some restrictive or inferior form of policy.

Implementation and review

Whatever form of commitments national governments adopt and in what order, negotiations must also consider how to ensure they follow through with the commitments they have made. Many environmentalists advocate 'treaties with teeth,' meaning commitments backed up by punitive sanctions for non-compliance, such as trade restrictions, fines, or withdrawals of aid or investment. But such sanctions are difficult to use effectively. They are hard to negotiate, and hard to agree to use in a specific case even once agreed in principle. They are often arbitrary or illegitimate in the way they are deployed, and often ineffective at changing the target country's behavior even when they are deployed. In view of these problems, international environmental agreements have usually avoided punitive sanctions, relying instead on softer methods of persuasion such as reporting and review processes, proceedings that focus more on collegial problem-solving than on identification and punishment of non-compliance, or other means of exerting moral pressure on national policy-makers. For the distinct problem of parties who want to meet their commitments but are unable to do so, these approaches can be augmented by various forms of financial and technical assistance.

How effectively commitments are implemented influences what commitments governments are willing to make, because what each government is willing to do depends on its confidence that others will fulfill their commitments. Consequently, procedures to monitor implementation deliver two types of incentives for governments to meet commitments: not just the risk of being embarrassed if they fail to meet their own, but also increased confidence that others are meeting theirs. Effective implementation procedures can consequently increase governments' willingness to agree to commitments, but only if these enjoy strong international support in principle. Strong implementation procedures cannot be used by a few activists to get the rest of the world, especially not the major powers, to participate in the first place if they are not interested in doing so.

Consequently, while negotiations have begun to develop procedures to review compliance with the climate regime, these are unlikely to make much contribution until a critical mass of major nations have adopted some form of substantive commitments that they clearly take seriously. Once that point is reached, there may develop increased willingness to accept stronger implementation measures in parallel with stronger commitments, because the more each nation relies on others to fulfill its commitments, the greater its interests in building a strong system of procedures and institutions for monitoring and enforcement, even if these apply to it as well.

In sum, the requirements of international climate policy, and its problems, in many respects mirror those of domestic policy. Reducing emissions requires incentives, which ideally should operate as uniform prices on emissions. To create such incentives, carbon taxes appear substantively preferable but cap-and-trade systems have the political momentum. Whatever policy commitment is chosen requires coordination, consistency, and transparency. The stringency of initial steps should be moderate, but with a gradual, pre-announced trajectory of increasing stringency over time to motivate the required long-term investment and technology development.

But international policy is harder to make, implement, and enforce than domestic policy, because the institutional structure and capacity for coordinated action is weaker. Consequently, making international climate policy requires even more compromises to satisfy political and administrative constraints than domestic policy. First and foremost, breaking the current deadlock of reciprocal demands that others take the first step, supported by opportunists on each side who do not want the deadlock to be broken, requires leadership. Leadership in this context means not just better proposals and arguments, but a willingness to demonstrate serious commitment to solving the problem by accepting costs and risks, even if the resultant unequal distribution of efforts means that first steps are not optimally cost-effective. Moreover, international policy-making cannot avoid some degree of explicit discussion of the distribution of burdens. In addition, international policy-making poses serious problems of implementation, in terms of both administration and confidence-building. Meeting these challenges will require innovation in ways to coordinate, calibrate, and link systems of national policy that are taking different approaches, including diverse mixes of taxes, cap-and-trade systems, regulations, and expenditures. It will also require willingness to compromise, informed by good practical judgment.

There are many ways that international policy can fall short of the ideal of maximum effectiveness and cost minimization. And some of these ways to get policy wrong may be so wrong-headed, costly, and ineffective that they would represent setbacks that make it harder to solve the climate problem. But other approaches that fall short, even far short, of the ideal may still represent first steps toward an effective approach that are much better than doing nothing. Approaches that involve a long period of preliminary measures which vary strongly in their stringency and seriousness, may be of this character – far from ideal, but a feasible step in the right direction, preferable to continued deadlock and inaction.

4.3 Putting it together: balancing benefits and costs of mitigation and adaptation

Up to this point, we have summarized current knowledge and uncertainty about the impacts of climate change, about potential responses for adaptation and mitigation, and about alternative ways to design policies to promote these. But in view of all this knowledge, how should we decide what to do? The most widely supported approach to evaluating complex public policy decisions is to examine their total social benefits, comparing the benefits of each proposed course of action to its costs. Preferred policies are those that make net social benefits – benefits minus costs – as large as possible. When a policy can be varied continuously over some range and the decision is at what level to set it – for example, how much to spend on health care and national defense, or how tightly to limit emissions of a pollutant – the level that maximizes net benefits is found by looking at marginal costs and benefits. As a pollutant is controlled more tightly, usually marginal costs increase and marginal benefits decrease: cutting the first ton gains a large benefit for a small cost, the second brings a little less benefit for a little more cost, and so on. Social benefits are maximized by controlling up to the last unit that brings a benefit larger than its cost, i.e. to the point where the marginal cost and marginal benefit of control are equal.

Because managing climate change requires choosing levels of two kinds of effort, mitigation and adaptation, maximizing social benefits in this case is a little more complex, requiring making three marginal quantities equal. Emissions should be cut until the marginal cost of mitigation – the cost of cutting one more ton – is equal to the marginal damage of climate change from the last ton emitted. But the damage of climate change can also be reduced by adaptation measures, which should be taken to the point where the marginal cost of the last increment of adaptation effort also equals the marginal climate damage avoided by that effort. So the ideal response sets the marginal costs of mitigation and adaptation, and the marginal damage from the remaining climate change, all equal.

This is fine in theory. But trying to identify optimal policy this way requires quantitative estimates of the costs of mitigation, the costs of adaptation measures, and the damage from climate change, over a wide range of possible mitigation and adaptation levels. We have some knowledge of all these quantities, and some insights into decisions that might be based on them, but these are all subject to substantial uncertainty and controversy.

4.3.1 Estimates of the cost of mitigation

Of the three types of cost, the most information is available on mitigation. Dozens of analyses of mitigation costs have been conducted, for the USA and other regions and for the world as a whole. These analyses use economic models that specify baseline assumptions of future population and economic growth, energy resources and markets, and technology trends. The models compare these baseline futures to alternative futures in which emissions are limited. For any specified emission limit, the models calculate the total and marginal cost of achieving the limit, relative to the assumed baseline. The marginal cost connects the analysis of emission limits and taxes; if a limit carries a specified marginal cost per ton, then that limit could in principle be achieved by a tax per ton equal to that marginal cost.

In all these mitigation-cost analyses, costs start low for small emission reductions. In the near-term (2030), the IPCC estimated that world emissions could be cut roughly 10 to 30 percent below baseline levels at costs below ~ $75 per ton carbon.[5] As the reductions increase, so does the cost: a 20 to 40 percent reduction could be made for less than ~ $180 per ton carbon. Other detailed technical analyses have found even larger opportunities for reduction, 30 to 45 percent below baseline at costs less than ~ $180 per ton carbon in the US in one study, and 70 percent below baseline at cost less than ~ $280 per ton carbon

[5] Recall that emissions, and prices of reducing them, can be expressed as either tons carbon or tons CO_2. This book expresses them as tons carbon, while the most recent IPCC assessment used tons CO_2. The two ways differ by the factor 44/12. So this marginal cost of $75 per ton C is roughly equal to $20 per ton CO_2, as stated in IPCC (2007c) on page 77.

worldwide in another.[6] In making these cost estimates, engineering-based models that aggregate the cost and contribution of specific technologies from the bottom up normally find more and cheaper reduction opportunities than top-down, economic-based models that infer reduction options from estimated substitution possibilities in the aggregate economy, but this difference has grown smaller as each type of model has been improved.

As larger emission cuts are considered, three things happen: costs increase; differences in cost estimates among models (an approximate but imperfect measure of uncertainty in the estimates) grow larger; and it becomes harder to meaningfully compare cost estimates among models because the assumptions they make on properties of the economy, the baseline, and the mitigation strategy vary on so many dimensions. Subject to those provisos, the IPCC summarized prior studies of the aggregate cost of various climate-stabilization levels. They found that emission trajectories aiming to stabilize around 535 to 590 ppm of CO_2-equivalent had total costs that ranged from 0.2 to 2.5 percent of world economic output in 2030, and from small benefits to 4 percent loss in 2050. Stricter stabilization targets, from 445 to 535 ppm CO_2-equivalent, had costs of up to 3 percent in 2030 and up to 5 percent in 2050.

The more recent analysis of stabilization scenarios conducted for the US Climate Change Science Program (CCSP) using three models considered joint controls of multiple greenhouse gases, including emissions related to land-use. In this analysis, the strictest target considered – forcing of 3.4 W/m^2, corresponding to about 520 ppm of total greenhouse gases expressed as CO_2-equivalent or 450 ppm of CO_2 alone – had costs equivalent to 1 to 2 percent of GDP in two models, and 6 to 16 percent in the third model. The second-tightest level – forcing of 4.7 W/m^2, equivalent to 660 ppm total CO_2-equivalent or 550 ppm of CO_2 alone – had costs of 0.3 to 0.8 percent GDP loss in two models, and 2 to 7 percent in the third. The marginal costs associated with these trajectories were, for the tightest trajectory, about \$170–190 (in two models) or \$384 (in one model) per ton carbon in 2030, about \$500 (in two models) or \$850 (in one model) in 2050; and for the second-tightest trajectory, \$13–26 or \$112 in 2030, \$36–70 or \$245 in 2050.

The wide variation of these estimates provides a reasonable picture of actual uncertainties, but does not mean we know nothing about mitigation costs. It is particularly instructive to look at what caused the variation in projected costs

[6] Translating emission prices into consumer price increases involves some approximation, because it depends on how final energy is produced from primary energy in the economy. A reasonable approximation is that a price of \$100 per ton carbon would raise the retail price of gasoline 30 cents per US gallon, of gas-generated electricity by ~ 1.5 cents per kWh, and of coal-generated electricity by ~ 3 cents per kWh (vs. current average electricity prices of about 10 cents/kWh in the USA).

among these models. In the CCSP study, the high costs in one model arose in part from higher baseline growth in GDP and emissions. But the largest source of variation lay in different assumptions of how readily the economy can substitute away from fossil-based energy, both through substitution of other inputs (capital and labor) and through innovations that lower the cost of climate-safe energy sources. The high-cost model assumes a stiffer, less substitutable economy, with less innovation in response to incentives generated by prices and mitigation policies. Costs in this model were further raised by strict, non-economic limits it imposed on expansion of nuclear energy. This key importance of technical innovation in determining mitigation costs makes sense. It could be inferred even in the IPCC SRES emissions scenarios, in the wide difference in emissions between the technological variants of the A1 scenario, even with the same population and economic growth assumptions. A more recent analysis posed the question more precisely by estimating the monetary benefit of early discovery of a cheap, zero-carbon energy technology. The value it found was more than three-quarters of the total baseline damages from climate change, a present-value of about $17 trillion.

How readily will technological innovation respond to incentives from mitigation policy? Evidence from other environmental issues suggests innovation is quite responsive, in that advance estimates of the cost of environmental improvements tend to be higher than the costs actually realized. This evidence would favor mitigation costs nearer the estimates of the two low-cost models rather than the high-cost one, but this remains a real uncertainty. Evidence from historical experience is probably not adequate to refine these cost estimates much further. On the other hand, these uncertainties also suggest that the rate of innovation, and its responsiveness to policy, might themselves be subject to influence by policy. Policies as discussed above, that put a price on emissions to give private actors incentives to commercialize emission-reducing innovations, could both advance knowledge about the responsiveness of innovation and make it more responsive. Similarly, increased public R&D investment in climate-safe technologies and support for technology assessment can help lower barriers to commercialization of emission-reducing innovations and bring costs down.

But while current cost estimates may be too high if climate-safe energy innovation is more responsive than they assume, they could also be too low because all these estimates assume optimal policy – equal marginal costs of emissions imposed everywhere in the world, with perfect and costless implementation, and any revenue the policies raise redistributed to the economy optimally. Policies that depart far from this ideal – by restricting flexibility in how nations and private actors cut their emissions, by allowing large and long-lived divergences in marginal costs, or by using revenues inefficiently – could sharply increase costs. In this sense, these differences in cost estimates do not so much reflect uncertainty, but rather provide guidance on how to design policies that achieve global mitigation at low cost.

4.3.2 Estimates of the cost of adaptation and climate-change impacts

In addition to mitigation costs, taking a cost-benefit approach to climate policy also requires estimates of the costs of climate-change impacts and of adaptation measures to reduce them. These are substantially more difficult to estimate than mitigation costs. Many studies have sought only to describe impacts, not evaluate or aggregate them. Many more have focused on particular impacts that are easy to assess, perhaps because they concern goods and services for which market prices provide good proxies for total social value (e.g., agricultural products, farmland, or coastal property), or perhaps because good data and models are available. For example, there are now many sophisticated studies of climate impacts on agriculture and impacts of sea-level rise, although even in these areas projected dollar values of impacts vary strongly from study to study.

Informing benefit-cost studies to guide climate policy decisions requires comprehensive assessments of all impacts, however – the easy and the hard to estimate, those for which markets exist and those for which they don't – as well as interactions among these. An example of a hard to assess, non-market impact might be the pleasure New Englanders derive from their climate and the landscape that depends on it, such as bright, snowy winter days, colorful fall foliage, and forests that support maple syrup production. Such impacts are hard to estimate for many reasons, including different valuations of particular aspects of climate among individuals and communities, and the possibility that people's preferences might shift over time. People might adapt to climate change as it occurs, learning to live with – or even like – the environment and climate they have, even if it means New England without snow or maple trees, as long as climate changes are slow enough for these preference changes to take place.

Studies have taken various approaches to assessing comprehensive impacts, none of them fully satisfactory. Some studies estimate the value of non-market effects by asking people how much they would be willing to pay to avoid them, but these methods are controversial. Some avoid assigning dollar values to all impacts by counting a few separate metrics of obvious importance. For example, one proposal counts five dimensions of impacts – market impacts, lives lost, biodiversity loss, income redistribution, and quality-of-life changes – but this approach may be both too complex to support benefit-cost assessments, and too simplistic to reflect all the important ways that people value climate, particularly in its quality-of-life effects. The studies that have presented aggregate, quantitative estimates of climate-change impacts have relied extensively on heuristic judgments, usually derived by sorting economic sectors and other domains of impact by presumed sensitivity to climate, assigning judgmental damage estimates in each domain, and adding them up. Moreover, these have all combined adaptation costs and damages into a single aggregate estimate, implicitly assuming optimal adaptation to any specified level of climate change.

Typical results from these studies have been that global-average warming of 2 to 3.5°C this century brings impacts equivalent to a loss of 1 to 3 percent in world economic output, but there is wide variation among studies. Although it appears clear that damages depend on the rate as well as the level of change, no comprehensive impact assessment has separated these. Other studies have estimated similar total losses – 1 to 2 percent of economic output – as the present value of discounted future losses over longer than this century, aggregating both smaller near-term damages and larger longer-term damages. A few analyses have found much larger damages from uncontrolled climate futures, such as 5 to 20 percent of economic output in the widely publicized UK report known as the "Stern Review." Estimates of the marginal social cost of emissions – how much damage does one additional ton emitted today inflict – also span a wide range. Many studies find this value between about \$35 and \$100–150 per ton carbon, while a few find values of more than \$1,000 and a few as low as \$3 – 5. The distribution of estimates is strongly skewed, with a long thin tail of very high estimates.

This wide range arises from many factors, including many specific assumptions about sensitivity of particular sectors to climate change, treatment of uncertainty, and how monetary estimates of damages are aggregated across world regions with widely varying incomes. But the largest source of the wide variation in these estimates is the discount rate, the amount by which future costs or benefits are reduced relative to present-day ones in aggregating to a single measure. Impacts of climate change grow larger over time in uncontrolled scenarios, so the largest damages appear late this century or beyond. Evaluated using a conventional approach such as a constant 5 to 10 percent annual discount rate, the present-value cost of these impacts becomes insignificant. When estimates of climate impacts are grouped according to the discount rate they use, those using the lowest rates, around 1 percent per year or lower, find the highest climate-change damages, in terms of both marginal and total costs. But with such low rates, such as the 1.4 percent rate used in the Stern Review, the largest share of calculated damages comes from projected impacts more than 200 years in the future. Such very low discount rates have the effect of making today's decisions highly sensitive to alternative assumptions about consequences of these decisions that occur centuries in the future. In contrast, studies using intermediate rates, such as the 3 to 5 percent often used for long-term public investments and policy decisions, calculate marginal social costs of emissions between roughly $35 and $100 per ton carbon-equivalent, and damages from uncontrolled climate change of roughly 1 to 3 percent of economic output.

There is no single right answer for the choice of a discount rate in climate change assessments. The framework for valuing effects over time in a growing economy comes from early twentieth-century economist Frank Ramsey, who identified that the discount rate depends not just on the normative choice of how to balance present and future welfare – trading off people's interests solely on the basis of when they live – but also on how fast incomes rise and how we trade off benefits to richer versus poorer people. Normally incomes grow over time, so future people are richer than present people. In this case, if we care more about the welfare of poorer people, this gives a higher weight to present people's welfare – not because they are present, but because they are poorer. The discount rate compounded from these two factors is the "right" one to use in analyses of long-term issues like climate change. Under certain assumptions, this discount rate equals the rate of return on investment, which unlike the two elements underlying discounting, can be observed.

While there is no correct value for the discount rate, there are two common errors in estimating it, one giving very high and the other very low rates. The first error treats the discount rate as identical to the normative choice of a trade-off between present and future people's welfare. Since many find it unacceptable to devalue people's welfare just because they live in the future, it is often proposed that this normative factor should be zero. But this does not mean the discount rate used to guide policy should be zero, because of economic growth. If economic growth makes future people richer, this may provide a proper basis to value monetary costs or benefits to them less, dollar for dollar, than costs or benefits borne by people today. The second error assumes the return we observe on investment – sometimes up to 10 percent – is the appropriate discount rate, ignoring the quite restrictive assumptions that are necessary for these to be equal.

4.3.3 Integrated assessment of climate-change adaptation and mitigation

Analyses that consider climate-change impacts, adaptation, and mitigation together are called integrated assessments (IA). Integrated assessment models represent the climate system, the socio-economic factors that drive emissions, the impacts of climate change, and potential mitigation and adaptation responses in a consistent quantitative framework. While highly simplified, these models can be used to simulate the effects of different mitigation and adaptation strategies. They can consequently be used to calculate the costs and benefits of alternative scenarios and policies, or to conduct the comparison of marginal benefits and marginal costs needed to identify optimal policies. In view of the complexity of combining all these components in a single model, each component is typically represented in highly simplified form. This is especially true for the representation of climate impacts. Because of the uncertainty of physically grounded impact projections, and the fact that they are available for only a few impact domains, integrated assessments have mostly relied on judgmental estimates of aggregate impacts expressed in monetary terms, as discussed above.

The earliest benefit-cost studies using integrated assessment models, conducted in the mid-1990s, tended to conclude that little mitigation was warranted. Optimal emission reductions this century were often only about 10 percent from the projected baseline. More recent studies have increasingly found more mitigation to be optimal, while retaining the shape of mitigation in which cuts diverge slowly from the baseline and grow progressively larger over time – the shape often called "the policy ramp."

For example, one recent benefit-cost analysis using the integrated-assessment model DICE found that the optimal emissions cut from the uncontrolled baseline trajectory was 15 percent immediately, rising to 25 percent in 2050 and 45 percent in 2100, to reduce warming in 2100 from 3.1°C to 2.6°C – substantial cuts, but on the low end of those currently proposed. The carbon tax required to produce this mitigation trajectory started at about $27 per ton carbon, rising consistently to reach $90 per ton in 2050 and $200 at the end of the century. Revised analyses with moderately strong climate constraints, limiting warming to 2.5°C or concentrations to 550 ppm CO_2-equivalent, required emission taxes only about 10 percent higher than these. Other benefit-cost analyses of optimal mitigation with different assumptions can yield sharply differing results, with optimal twenty-first-century emissions reduction varying from as little as 10 percent to more than 80 percent, and associated emission taxes varying more than ten-fold, with the largest variation occurring later in the century.

The sources of these large differences are principally factors we have already discussed, and their interactions: how future effects are discounted, alternative assumptions about the determinants of technological change, and the treatment of uncertainty, particularly the risk of extreme, catastrophic, or irreversible changes. Because changes in emissions only make significant change in climate some decades later, greater discounting of future effects leads to less near-term mitigation, and generally favors adaptation – and bearing the resultant harms – over mitigation. Assuming technological change responds strongly to policies and prices tends to favor more early mitigation, while assuming more rapid technological change independent of policies and prices tends to favor less early mitigation. Uncertainty in climate impacts, particularly the possibility of extreme impacts, favors stronger early mitigation if we are risk-averse. We are frequently risk-averse in policy, defending against risks with relatively low probability if their consequences are bad enough, for example in defense and public health policy. But the effect of uncertainty also depends on the source of the uncertainty. If, as in some analyses, future economic growth contributes much of the uncertainty in future climate change, this would favor less costly early mitigation.

Integrated-assessment models provide a valuable framework for thinking through the large-scale structure of climate response, for evaluating how to balance mitigation and adaptation efforts over time, and for identifying key uncertainties for research. Moreover, while different assumptions can produce fairly wide variation in preferred policies, there is substantial convergence developing on several key points. All analyses identify benefits from emission reductions. Although the preferred reductions vary substantially, they grow large by late in the century in all analyses. Moreover, all analyses favor some form of graduated policy, in which the stringency of control measures and the associated price on emissions start at a modest but non-trivial level – most analyses recommend initial emission prices from about $30 to $60 per ton carbon – but with a sustained, long-term increase, so emission prices reach hundreds of dollars by late in the century. Beyond these points of convergence, particular analyses show substantial variation, particularly later in the century, depending principally on alternative assumptions about uncertainty and risk-aversion, technological change, and discounting. Some of these are uncertainties about properties of the economy or climate system that, while not justifying delay in starting action, will be priorities for further research to refine understanding of preferred policy courses over time. Others, such as alternative approaches to discounting, at least partly reflect different normative presumptions about valuing future effects, which are not amenable to resolution through research.

4.4 The third class of response: geoengineering

In addition to mitigation and adaptation, a third class of potential responses to the climate issue involves actively manipulating the climate to offset the effects of increased greenhouse gases in the atmosphere. This approach, usually called geoengineering, includes a diverse collection of proposals. Some involve obstructing incoming sunlight, by injecting reflective aerosols into the stratosphere or launching screens into space to block a little of the Sun's light from reaching the Earth. Others involve manipulating the global carbon cycle, for example increasing ocean CO_2 uptake by fertilizing marine plankton with some limiting nutrient such as iron, or directly removing CO_2 from the atmosphere. Several geoengineering approaches appear highly promising in early studies, with early cost estimates substantially lower than conventional mitigation.

The greatest potential value of geoengineering is that some approaches can modify the climate much faster than mitigation can. While even an extreme program of mitigation would take decades to begin slowing climate change, a Sun shield in space would offset the effect of elevated CO_2 as soon as it was deployed, which might take a decade or so. A program to inject aerosols in the stratosphere could be deployed even faster. Consequently, geoengineering approaches could provide protection against worst-case scenarios in which we badly fail to limit climate change or are unlucky in how rapid and severe it turns out to be. The scale of effort required could likely be achieved by one or a few rich and technically advanced nations.

Such active manipulation of the Earth on planetary scale carries large risks, however. Some potential mechanisms of environmental disruption have already been identified, such as disruption of the ozone layer from stratospheric aerosol injection, but there may be other environmental risks not yet anticipated. Still other risks may arise through how the prospect of cheap geoengineering may influence other decisions to manage climate-change risks. Geoengineering does not necessarily offset all environmental harms from greenhouse gases. For example, offsetting the radiative effect of greenhouse gases through stratospheric aerosol injection would not reduce the direct effects of elevated atmospheric CO_2, such as ecosystem change and ocean acidification. Yet even if geoengineering should turn out to be inadequate, too risky, or politically unacceptable, the prospect it provides of a cheap technical solution may undercut early efforts at mitigation that in retrospect turn out to have been necessary.

Geoengineering projects may also pose serious legal, diplomatic, and political problems, such as conflict over who has the authority to undertake them, whether they would impinge on existing international treaties or require new ones, and how control over geoengineering and the burden of funding it would be shared. Most problematic would be the potential for conflict between nations proposing such a project and those opposing it, either because of principled opposition to active planetary-scale manipulation or because the opponents expect the climatic effects of the project to harm them. In the extreme, a geoengineering project could be regarded as a hostile act by another state, akin to Cold War proposals to use active weather modification as a weapon.

It is clear that present efforts are far from exhausting attractive opportunities for mitigation and adaptation, so there is no basis for expecting geoengineering to play any significant near-term role in the response to climate change. Still, despite their large evident challenges, these approaches merit serious consideration. Further research and development and comprehensive careful assessment of risks are clearly warranted so options can be available in the event they are needed, even if no near-term deployment is expected. If things are looking dire in 2030 or 2050, a technical response to arrest or reverse some changes within another decade or so may be the least bad option then available – despite its risks, and despite the severe political and legal problems it would pose.

4.5 Conclusion: policy choices under uncertainty

The most basic question in choosing a response to climate change is how to act responsibly under uncertainty. Although uncertainties in climate change are not as overwhelming and debilitating as some advocates suggest, they still pervade every part of the problem. Future paths of CO_2 emissions have wide uncertainty. So do projections of the global climate response and the resultant regional-scale impacts. So too do estimates of the efficacy, costs, and other effects of the various approaches to mitigation and adaptation.

About some of these matters we know quite a lot, about others we know less, and about some we know very little. Some of these uncertainties derive from limits in our knowledge of Earth systems. Much uncertainty also derives from our limited ability to predict human behavior and patterns of development. While progress in knowledge can be expected, neither of these types of uncertainty will be reduced to insignificance anytime soon.

Yet despite the existence and persistence of these uncertainties, there are several points about responding to climate change that are powerfully supported by current knowledge. Evidence of the reality of climate change and the risks it poses has far passed the threshold at which economic analysis and reasonable prudence call for a response. Responses that rationally balance costs and benefits combine adaptation to reduce harms from climate change with mitigation to slow and stop climate change, as well as further research and assessment of geoengineering options as a stopgap against potential failure of these approaches. Adaptation is necessary because we cannot stop climate change anytime soon. Effective adaptation will require changes by many actors who are not yet accustomed to considering climate change in their planning. Adaptation can be enhanced by government policies and expenditures, and is likely to have substantial distributive implications since in many cases poor people, communities, and regions are likely to be most vulnerable to climate-change impacts.

Mitigation is necessary to limit the climate changes to which we must adapt, and must begin early to reduce reliance on adaptation later. Many analyses now give similar advice regarding the broad shape of near-term mitigation strategies, of which the central component is a set of market-based policies that put a price on emissions, starting modestly and increasing over time.

Yet this convergence of judgments on some elements of near-term action does not mean that uncertainties have gone away or are not important. On the contrary, as in any high-stakes domain of human affairs, choices must be made despite continuing uncertainty, and responsible choices require balancing prudent near-term actions, continuing efforts to learn more, and provisions to adapt and revise decisions in response to experience, new knowledge, and changed capabilities. As a general matter, these principles of decision-making under uncertainty are widely accepted, but how to apply them to the climate-change issue has been contentious in the extreme. In the next and final chapter, we discuss the political debate over climate change policy as well as our recommendations for a responsible and practical path forward for responding to the climate-change issue.

Further reading for Chapter 4

Leon Clarke, J. Edmonds, H. Jacoby, H. Pitcher, J. Reilly, and R. Richels (2007). *Scenarios of Greenhouse Gas Emissions and Atmospheric Concentrations*. Synthesis and Assessment Product 2.1a. Washington, DC: US Climate Change Science Program.

A recent analysis of the economic, technological, and policy implications of stabilizing climate change at four alternative levels, based on joint controls of multiple greenhouse gases, using three integrated-assessment models.

IPCC (2000). *Emission Scenarios*. Special report of the Intergovernmental Panel on Climate Change. N. Nakicenovic and R. Swart (eds.). Cambridge, UK: Cambridge University Press.

IPCC (2007b). *Climate Change 2007: Impacts, Adaptation and Vulnerability*. Contribution of Working Group II to the Fourth Assessment Report of the Intergovernmental Panel on Climate Change, M. L. Parry, O. F. Canziani, J. P. Palutikof, P. J. van der Linden, and C. E. Hanson (eds.). Cambridge, UK: Cambridge University Press.

This is the most recent full assessment of the IPCC's Working Group II, which reviews knowledge of potential impacts of climate change, ability to adapt, and vulnerability of environmental and social systems to climate change.

IPCC (2007c). *Climate Change 2007: Mitigation of Climate Change*. Contribution of Working Group III to the Fourth Assessment Report of the Intergovernmental Panel on Climate Change, B. Metz, O. Davidson, P. Bosch, R. Dave, and L. Meyer (eds.). Cambridge, UK: Cambridge University Press.

This is the most recent full assessment of the IPCC's Working Group III, which reviews knowledge of technical and economic opportunities to reduce net emissions of greenhouse gases and policy instruments to promote such reductions.

IPCC (2007d). *Climate Change 2007: Synthesis Report*. Core Writing Team, R.K., Pachauri, and A. Reisinger (eds.). Cambridge, UK: Cambridge University Press.

The IPCC "synthesis report" combines results from an assessment's three working groups to present an integrated analysis of the implications of alternative emissions trajectories and reduction goals.

Richard Moss *et al* (2009). *Representative Concentration Pathways: A New Approach to Scenario Development for the IPCC. Nature*, in press. Available at: www.pnl.gov/gtsp/publications/2008/papers/20080903_nature_rcp_new_ scenarios.pdf

The 2000 special report on emissions scenarios provided the results and background for the IPCC baseline emissions scenarios used as inputs to climate-model projections through the IPCC Fourth Assessment Report. The Moss et al. paper outlines the new process for generating climate scenarios, starting with coordination on alternative pathways for radiative forcing, to be used in the IPCC Fifth Assessment Report.

William Nordhaus (2008). *A Question of Balance: Weighing the Options on Global Warming Policies*. New Haven: Yale University Press.

The most recent analysis of climate mitigation and adaptation responses, using the DICE integrated-assessment model. Includes a discussion of the Stern Review and the role of discounting assumptions in generating its large climate-damage estimates, as well as a discussion of the policy choice between emission taxes and cap-and-trade systems.

Nicholas H. Stern (2007). *The Economics of Climate Change: The Stern Review*. Cambridge, UK: Cambridge University Press.

A comprehensive economic assessment of climate change, led by the former chief economic advisor of the UK government. Controversy over this assessment's high climate impact estimates, mainly due to a low discount rate, generated substantially increased attention to the problem of identifying socially optimal climate strategies.

Jefferson W. Tester, E.M. Drake, M.J. Driscoll, M.W. Golay, and W.A. Paters (2005). *Sustainable Energy: Choosing Among Options*. Cambridge, MA: MIT Press.

A comprehensive text on alternative energy resources and technologies, and the key uncertainties and analytic methods involved in assessing alternative energy futures.

5

The state of climate policy and a path forward

The previous chapters have summarized present knowledge and uncertainty on the climate and how it is changing, the evidence for human causes of the observed changes, and the range of changes projected this century, as well as the less definitive current knowledge about potential climate change impacts and responses. This final chapter is more political, in a few senses. First, we examine the present politics of the issue, reviewing major actors' policies and positions. Second, we summarize and assess the main arguments being made against serious action to limit climate change. Finally, we present our own judgments of what kind of response to the climate issue appears appropriate in view of present scientific knowledge and political possibilities.

5.1 Climate-change politics: current policies and positions

Although climate change came onto policy agendas as early as 1990, little progress was made in policy debates in the 1990s and virtually none between 2001 and 2008, either in the United States or at the international level. During this period, the European Union and a few of its leading member states, as well as a few North American States and Provinces, took significant initial steps, but even these fell far short of what is needed to start the required energy-sector transformation. Other jurisdictions took only symbolic and small actions vastly too weak for the job, or none at all.

This situation began changing in late 2007 and 2008, when several factors – including new scientific results and assessments, heightened public concern in many nations, increasing activism in the US congress and state governments, and the 2008 US election campaign – combined to raise support for near-term action. In the aftermath of the Copenhagen meeting, the policy proposals and arguments being advanced on climate change are a rapidly moving target, so any characterization we make here risks being quickly outdated. Still, to provide context for continuing policy debates, this section presents a snapshot of where policy proposals and debates stand in early 2010.

As Chapter 4 discussed, the major responses to climate change include mitigation measures to reduce greenhouse gas emissions, adaptation measures to reduce vulnerability to climate-change impacts, and geoengineering options to be investigated as insurance against the risk of unexpectedly severe climate change or failure of mitigation and adaptation approaches. But current debate is mainly focused on mitigation, because it is the response for which long lags most clearly require near-term decisions, and the response that most clearly involves exercise of government authority over private actions. With a few exceptions, adaptation and geoengineering are perceived as posing future choices rather than immediate ones, and as less intrusive into citizens' lives and choices. Here, we follow the priorities of the current debate: while noting that adaptation will be an essential part of climate-change strategy, and geoengineering measures may also turn out to be important, we focus primarily on decisions and debates about mitigation.

In the USA, climate policy stagnated for several years after President Bush's 2001 decision not to ratify the Kyoto Protocol. As Chapter 1 summarized, US policies under the Bush administration included only a weak, non-binding target for emissions intensity (emissions per dollar of GDP), plus voluntary programs and research. Starting in 2003, however, congressional bills began proposing stronger measures, including nationwide emissions targets and tradable emission permit systems for the largest point sources. Congressional activity grew through 2008, when seven climate-change bills were introduced, proposing emission cuts as deep as 75 percent by 2050. But while these bills represented an advance in the debate, none of them was enacted.

With little action at the federal level, US climate leadership passed to the States around 2002. Most states have now enacted some form of climate policy or action plan, and more than 20 have adopted comprehensive reduction targets for statewide emissions. In addition, three inter-state groups – in the west, midwest, and northeast – are developing regional cap-and-trade systems for large emission sources. The northeast group, the Regional Greenhouse Gas Initiative (RGGI), began auctions in 2008 for permits required of major sources starting in 2009. The high cap in this system produced excess supply of permits, however – especially in the 2008–2009 recession – so the price of these permits has been only \$10 to \$15 per ton C (\$3 to \$4/t CO_2).

The strongest state action has been in California, which has enacted a mandatory limit on state emissions at 1990 levels by 2020, with an additional goal of cutting 80 percent by 2050; standards for motor vehicles, appliances, and the carbon content of fuels; and a cap-and-trade system under development to integrate with the Western Climate Initiative's regional program. The vehicle standards, which rely on California's unique authority under the US Clean Air Act to regulate motor vehicle pollution, require phased reductions in vehicle CO_2 emissions, starting in 2009 and reaching 30 percent cuts by 2016. This standard was initially delayed by legal challenges and by EPA rejection of the Clean Air Act waiver that California requires to regulate beyond national standards. The Obama administration reversed this rejection, then in May 2009 announced stringent new nationwide standards similar to those proposed by California, requiring average fuel economy of 35.5 miles per gallon for cars and light trucks by 2016.

Intensive work continued through 2009 to develop greenhouse-gas policies in both Congress and the Executive branch. A major climate and energy bill, which is broadly consistent with statements from the Administration, was narrowly enacted by the House of Representatives in June. This bill includes a series of targets for reducing US emissions, increasing over time to reach roughly 80 percent reduction by 2050. These targets are to be achieved mainly by a cap-and-trade system with the auctioned share increasing over time, and revenues used to fund clean energy technologies and tax credits for workers. The bill also includes provisions for sectoral emissions regulations, support for research and development (R&D), and processes for scientific and technical assessment and review of regulatory measures. Specific policy provisions will remain in flux as the legislative process continues, beginning with action by the Senate, where a broadly similar bill was voted out by the Environment and Public Works Committee in November 2009 over the objections of the Republican minority on the Committee, but has not yet been brought to the Senate floor.

While supporting development of new legislation, the Executive branch is also responding to a 2007 Supreme Court decision by preparing to regulate greenhouse gases under the existing Clean Air Act. In *Massachusetts v.* EPA, the Court found that greenhouse gases are pollutants as defined in the Act, so EPA must issue a so-called "endangerment finding" on whether they endanger public health or welfare, and must regulate them if they do. EPA issued the required finding in March 2009, and proposed regulations of emissions from large stationary sources and vehicles in September. Greenhouse-gas regulation would fit awkwardly into the framework of the Clean Air Act, so it would be preferable to enact new legislation tailored for this purposed. But at a minimum, this clear establishment of EPA's legislative authority can provide a useful threat to motivate political support for better crafted, more efficient greenhouse-gas controls under new legislation. At the same time, some legislative proposals in circulation have removed this Clean Air Act authority, so the new legislation would provide the only statutory authority to control greenhouse gas emissions.

These negotiations will in all likelihood continue until the current debate over new climate legislation is completed.

In contrast to the US, where development of climate-change policy suddenly became more serious during the 2008 election campaign, the European Union has pursued ambitious climate initiatives for more than a decade. Some EU states are likely to meet their first-period Kyoto commitments, although recent data suggest the EU as a whole will fall somewhat short. In the strongest positions are the UK and Germany, which will both reduce emissions well beyond their Kyoto commitments due to a combination of fortunate circumstances and strong policies. The EU also has the strongest commitments in place to control future emissions. The most recent European climate plan – enacted in December 2008, over substantial resistance from industry groups and some member-states – re-affirmed the existing EU target to cut emissions 20 percent below 1990 levels by 2020, or up to 30 percent if other major industrial countries enact similar restrictions. The program includes standards for renewable electricity, vehicle efficiency, and low-carbon fuel, national emissions caps for specified sectors, a fund for carbon-capture demonstration projects, and other measures.

The centerpiece of the EU climate program is an emissions trading system for large sources, covering about 40 percent of total emissions. The first phase of this system, from 2005 to 2007, suffered from high permit allocations by national authorities and experienced high price volatility, including a sudden price collapse in mid-2006. The second phase, matching the 2008–2012 Kyoto commitment period, has seen more conflict but more achievement, as the EU Commission required several nations to reduce their allocations to closer approach the overall Kyoto target. In both these phases, the system required that nearly all permits be distributed for free. Plans to move toward auctioning in the third phase, from 2013 to 2017, met forceful opposition, and a compromise that reduced initial auctioning to 20 percent was necessary to secure adoption of the new Climate Plan. Negotiations are still underway to expand the system to include emissions from aviation and shipping, and to implement a border carbon tax to limit competitive losses if the EU's trading partners fail to take similarly stringent action.

Other industrialized countries have enacted programs to limit emissions, of varying degrees of seriousness, but none sufficient to meet their Kyoto commitments. Japan's 2005 climate program included sectoral reduction targets, building efficiency standards, promotion of low-emissions vehicles through government purchasing programs, as well as various voluntary measures and substantial reliance on sinks and purchased credits. Additional pledges made in 2008 and after a change of government in 2009 included enacting a nationwide cap-and-trade program, and committing to an aggregate emissions reduction of

25% below 1990 levels by 2020. Of the major industrial-country parties, Canada is falling farthest short of its Kyoto commitment, having ratified late after a decade of strong emission growth, weak mitigation programs, and unproductive consultations seeking consensus on mitigation. As in the US, some Canadian Provinces have taken stronger steps than the national government: British Columbia leads the way with a comprehensive program, including a provincial carbon tax.

The past few years have also seen remarkable development of greenhouse-gas plans and policies in developing countries and emerging economies. Thus far, only two major developing economies have stated targets for total national emissions. South Korea has pledged to cut emissions 4 percent below 2005 levels by 2020, supported by multiple policies including a cap-and-trade system and an innovative program for greenhouse-gas footprint labling of products. South Africa has stated a plan for emissions to peak by 2025, stabilize for ten years, then decline. Many other developing countries have enacted mitigation policies, and several have announced targets to reduce emissions intensity (emissions per GDP). In late 2009, China committed to reduce intensity 40–45 percent below 2005 levels by 2020, while India announced it would reduce intensity 24 percent by the same date. Both targets are supported by multiple domestic laws and aggressive clean-energy innovation strategies, especially in China. One NGO projected that policies already enacted will reduce 2020 emissions below baseline 13 percent in China and 19 percent in India, despite projected rapid economic growth.

These national initiatives are all taking place in the shadow of continuing international negotiations. As Chapter 1 discussed, the Kyoto Protocol entered into force in 2005, following Russian ratification, and virtually all nations except the United States are now parties. Kyoto's 2008–2012 first commitment period is now underway, but very few nations that accepted national emissions limits for this period (called "Annex 1 nations") will achieve them.

Despite these decisions, however, negotiations since Bali have achieved little advance from the prior deadlock. Little or no progress has been achieved on the most important agenda items, the nature of new mitigation commitments for industrialized and developing countries and the relationship between them, and the nature of technology and finance initiatives to support developing-country participation. Mitigation negotiations for industrialized and developing countries have proceeded on separate tracks, and developing countries have insisted on keeping these unlinked and settling industrialized-country commitments first. Within industrialized-country discussions in December 2008, European proposals for 25 to 40 percent cuts by 2020 met strong resistance from Japan, Russia, Canada, Australia and others – all but the United States, which was in the transition between administrations and not participating actively. The largest achievement thus far has been agreement on an "adaptation fund" to be financed by a levy on international mitigation projects, but the sums involved are many orders smaller than the tens of billions each year needed to support developing-country mitigation and adaptation. Many participants are now focused on lowering expectations for Copenhagen – as has been done many times in the history of international climate-change efforts – re-defining its purpose as starting and structuring serious negotiations rather than completing them. A major US initiative could energize the negotiations, but probably not in time to yield concrete results in Copenhagen.

Among major non-government actors, environmental groups favor early mitigation and most support the strictest targets (e.g., stabilizing concentrations at 450 ppm CO_2-equivalent, with increasing attention to even tighter targets such as 350 ppm), and the strictest requirements for near-term action (e.g., the upper endpoint of the EU position, 40 percent cuts for industrialized countries by 2020). Industry positions on mitigation are more mixed, and have shown more substantial movement over the past few years. In early climate debates in the 1990s, most industry actors strongly opposed mitigation, and many supported industry-based NGOs whose opposition extended to denying scientific evidence. The most prominent such group was the Global Climate Coalition, formed in 1989 by a group of fossil-fuel producers and major energy-using industrial firms. This coalition began unraveling in the late 1990s as members rejected its extreme stance, beginning with Du Pont and BP in 1997, and Royal Dutch/Shell in 1998. The loss of two major oil producers was especially significant, as these firms began re-positioning themselves as environmentally responsible energy companies.

A few business sectors began supporting mitigation early. These included minor groups like the skiing industry (the 'small island states' of the private sector), which endorsed a mitigation bill in 2003, as well as major sectors like insurance and finance, which view climate change as both a risk to asset values and a market opportunity for new financial instruments. At first these industry groups carried less clout than those opposing mitigation or sitting on the fence, but this balance has shifted over time. An increasing number of major firms have now accepted the seriousness of climate change and shifted from simply opposing mitigation to advocating principles for mitigation policies they can live with – e.g., that policies must be cost-effective, international, based on scientific assessment of risks, and applied broadly and equitably rather than targeting particular industries. By 2005, many energy producers, electrical utilities, and industrial firms had partnered with environmental NGOs to support orderly emission cuts. One such group, the US Climate Action Partnership, called for a US cap-and-trade system in 2007 and proposed a detailed policy blueprint in early 2009, including phased cuts reaching 80 percent by 2050 – similar to current administration and Congressional proposals.

As always in large and diverse groups, this shift in industry views toward cautious support for mitigation reflects diverse motives and interests. Leaders of some firms may have sincerely changed their views on the scientific evidence for serious climate-change risks. Others may judge that mitigation is inevitable sooner or later, so it is prudent to engage negotiations constructively. Still others may judge that the shift to climate-safe energy technologies holds business opportunities. And some may be maneuvering to capture the rents that will no doubt be conferred by strict mitigation policies.

This shift in industry positions has left a group diminished in numbers and stature still firmly opposing action on climate change, also based on diverse reasons and interests. Of these, some perceive their core interests are threatened by any serious mitigation effort. For example, remaining opponents include some heavily invested in coal or heavy hydrocarbons such as oil sands, although even among these some firms are hedging their bets and counting on carbon capture – promoted under the slogan "clean coal" – to sustain them. Other organizations and individuals perceive climate change and potential responses mainly in terms of contending political ideologies, so evidence of the scientific basis for climate-change risks or technological prospects for climate-safe energy are unlikely to persuade them. With this range of views, the political landscape for climate-change action has clearly shifted, but remains complex and contested.

5.2 Climate-change politics: remaining arguments against action

In view of the recent shift of political views to heightened concern about climate-change risks and increased support for serious action, it would be reasonable to expect that the debate had shifted from whether to do anything, to what to do. But this is not entirely correct. The politics of climate action remains contentious, with many actors and organizations still forcefully opposing anything beyond the most trivial near-term action. Although reduced in numbers, stature, and influence, this group still includes many prominent figures in politics and government, as well as some industry leaders and many NGOs and individuals. These groups are diverse in their reasons for opposing action, but there is significant commonality in the claims they make to support inaction. As scientific evidence for climate change and political support for action strengthen, some arguments being advanced are becoming increasingly extreme, shrill, and implausible. But these arguments are still espoused by powerful political actors and exercise continuing influence over public opinion, so must be addressed. This section does so.

Like current policy positions, the claims being made to oppose climate action are a moving target. Still, there is some continuity, both over time and from issue to issue, so it is useful to provide a snapshot. We gather these arguments into four groups: attacks on the Kyoto Protocol; claims that serious mitigation will be ruinously costly; denial of scientific evidence for climate change and related attacks on scientific processes and institutions; and generic arguments about uncertainty and its significance for action or inaction. This section summarizes and critiques these arguments in turn.

A final type of claim, which is growing more prominent as other arguments become increasingly untenable, relies on pure rhetorical devices with no substantive basis. For example, advocates increasingly frame the climate issue in the form of the question, "Is climate change a crisis?" The response to this depends predominantly on the definitional question of what constitutes a crisis, terrain where the advantage lies with the more skilled debater, not the one with stronger scientific evidence to offer about climate change risks. These debating tactics frequently invoke broad political framings or symbols, treating climate change as a conflict of political ideologies rather than an issue that turns on assessment of scientific evidence or judgments of what action that evidence warrants. They are also frequently coupled with personal attacks on public figures who advocate strong action, such as former Vice-President Al Gore or NASA scientist James Hansen. These tactics can be powerful in public debates, particularly in mobilizing support based on pre-existing political affinities, but they have no substantive content so we do not address them here.

Attacks on the Kyoto Protocol

Although the Kyoto Protocol and its weaknesses are of decreasing importance as the focus of policy-making and negotiation moves beyond the end of the Kyoto commitment period in 2012, attacks on the Protocol still occur frequently in arguments against mitigation. For several years after its adoption, the Protocol was often attacked as "fatally flawed." Many of these attacks became irrelevant as subsequent negotiations corrected the Protocol's most acute weaknesses, but three critiques against its core mitigation commitments have persisted: that they require no participation from major developing countries; that they are arbitrary and not based on science; and that they are simultaneously too strong and too week – too strong and thus too costly in the near term, but too weak to achieve significant long-term reduction of global climate change.

Of these criticisms, the first two are factually correct but never sufficed to make the case for rejecting the Protocol. As Chapter 4 discussed, finding an acceptable basis for sharing efforts among nations is one of the greatest challenges in climate-change negotiations. The Kyoto Protocol imposed near-term emissions targets only on industrialized countries for both practical and principled reasons, but only did this as a first step. It was widely understood that developing countries' responsibility to share in mitigation efforts would grow as their economies and emissions grew – the question was when, and how much. This remains the most important and contentious point of current negotiations.

It is also correct that the Protocol's emission limits are arbitrary, because they were a bargained compromise between some nations that sought stricter targets and others that sought weaker ones or none. In this, they resemble all politically negotiated outcomes: being arbitrary is no special weakness of the Kyoto Protocol. Nor are the targets "based on science," because scientific knowledge cannot specify any particular target. Science can inform decisions about targets, by projecting the consequences of alternative emission paths – faster climate change from weak emission controls, slower change from strong ones. Science might even identify emission paths carrying increased risk of abrupt climate changes, although this would require substantial advances from present knowledge. But without some confidently known environmental threshold that all agree must be avoided, no emission target is more or less 'based on science' than any other. This charge can be made against any target, but is essentially meaningless.

The third charge made against the Protocol's targets – that they are both too strong and too weak – has changed with the passage of time since the Protocol was adopted in 1997. At that time, with more than a decade of lead time to achieve the targets, they appeared ambitious but achievable. But as years passed without serious mitigation efforts and emissions continued to grow, the targets became increasingly costly and eventually unattainable. At the same time, the targets only fixed emissions over one near-term, five-year period, while stabilization requires emissions to decline sharply over several decades. So the targets were indeed too weak to stabilize climate change, because they were not intended to be a complete solution, but a first step to be strengthened over time. This is how effective environmental treaties work.

These attacks on Kyoto are now overtaken by events, as negotiators work toward a new climate agreement. Yet they remain relevant because similar criticisms can, and likely will, be leveled against any future agreement that includes emissions targets. The specific terms of the next climate agreement will again reflect political bargaining among actors with different views of how much mitigation the evidence justifies. Consequently, they will once again be arbitrary, political, and not based on science, just like the Kyoto targets. Similarly, the next climate agreement will again be just a next step – hopefully a more serious one – that will impose near-term costs, yet be too weak by itself to stabilize climate. Consequently, it will be vulnerable to the same attacks made against the Kyoto Protocol: by ignoring the evolutionary character of any environmental agreement, treating it as a once-and-for-all solution rather than one step on an imperfectly perceived path toward the required large-scale changes; and by comparing it to a presently unattainable ideal solution, instead of to the alternative choices reasonably available. These rhetorical attacks will remain available and can be anticipated, but this does not mean they have any merit.

Mitigation will be ruinously costly

A second set of arguments for inaction claim that a serious mitigation effort would be so costly as to be economically ruinous. Some of these arguments use biased assumptions to exaggerate current cost estimates, while others focus only on the costs of mitigation without considering the benefits that mitigation brings by avoiding risks of climate change.

We discussed current estimates of mitigation costs in Chapter 4. Mitigation of the scale needed to stabilize climate at reasonably prudent levels will be costly, although the most credible analyses suggest costs will be manageable – a percent or two loss of future economic output. Many technologies to reduce emissions are available or in development, offering a large menu of possibilities stretching smoothly from the small and immediate to the large and long-term. The diversity of technological options available is crucial to our confidence that costs will be manageable, because it means the feasibility of large reductions does not depend on any one technology. Costs will not be equally distributed across the economy, however: indeed, sectors likely to bear higher costs from the shift toward climate-safe energy have an interest in exaggerating the cost to society overall. But high sectoral costs represent a political problem of finding an acceptable distribution of costs, not a reason to avoid acting at all.

Mitigation costs also have substantial uncertainty, as the studies summarized in Chapter 4 show. Consequently, it is easy to generate much higher cost estimates, e.g., by considering a limited set of technologies, by selectively focusing on faults and limitations of each technology (particularly immature ones), or by ignoring innovation and instead considering only technologies that are commercial today. It is also easy to produce high cost estimates by assuming badly designed mitigation policies. But to the extent that cost uncertainties depend ails of policy, this provides guidance on how to design good policies. And xtent that cost uncertainty depends on economic and technological operties that are presently unknown, there now appears to be little prospect of further refining cost estimates without starting the effort. Having any hope of achieving the required reductions requires starting immediately despite unavoidable uncertainties, then adapting decisions and policies as we learn more.

Finally, no mitigation cost is warranted if mitigation brings no benefits. Consequently, it is impossible to assess whether some level of mitigation effort costs too much without also considering what benefits it provides. Claims that the costs of limiting climate change are too large are thus often linked with claims that the benefits of doing so are small, i.e., claims about the severity of climate change and the impacts that would be avoided by mitigation. These in turn are linked to the most common arguments against action on climate change, those that deny the scientific basis for the risk.

Scientific denial revisited: attacking the IPCC

Of all arguments made against action on climate change, the most prominent target climate science, either denying particular points of scientific knowledge or attacking the integrity of the scientists and scientific assessment processes that communicate that knowledge. Section 3.5 summarized some of the common claims advanced in policy arenas that deny specific points of climate-science knowledge, and outlined why these are wrong. This section examines the higher-level rhetorical strategy in which these specific claims fit, and the attacks on scientists and scientific assessment bodies that increasingly accompany them.

The specific claims that deny the reality and seriousness of climate change follow a four-part rhetorical structure, of a form that lawyers call "arguing in the alternative." These claims state in the first place that climate change is not happening; second, that even if it is happening, it is of natural, not human, cause; third, that even if it is human-caused, it will not change much more, but will be self-limiting due to some form of negative feedback; and fourth, that even if the changes are large, they will on balance not be particularly harmful and may even be beneficial.

Similar claims denying scientific knowledge with the same rhetorical structure have been advanced repeatedly on other environmental issues, whenever these have produced contested policy debates. Paradoxically, such claims often grow more prominent as scientific knowledge strengthens. Like negative political campaigns, misrepresenting scientific knowledge in policy debates is a high-risk but potentially effective strategy available to a side that is losing. The ozone layer again provides a close and instructive parallel. In the early 1990s, as policy consensus developed to phase out ozone-depleting chemicals, based on strongly converging – although not complete or perfect – scientific knowledge about their contribution to ozone depletion, a fierce backlash developed that widely circulated in policy debates several real remaining scientific uncertainties and anomalies, together with a diverse group of long-refuted and ridiculous claims, in an attempt to resist the growing policy consensus. Participants in this ozone backlash – including several individuals now playing the same role in the climate debate – asserted that ozone was not decreasing, that any observed decreases were due to natural fluctuations, that future losses would be self-limiting, and that the impacts of ozone loss would be modest or even beneficial, but were overstated by alarmists and opportunists. Similarly on climate change, as the evidence for the reality and seriousness of climate change has grown so strong that most citizens and policy-makers judge a serious response to be warranted, those who wish to use scientific claims to oppose action have been forced to resort to increasingly biased and misleading claims, or outright false ones.

Advocates use these tactics of misrepresenting scientific knowledge, or obscuring or denying settled questions, because they often work. Claims called "scientific" get special respect in policy debates and often succeed at persuading, particularly when they support listeners' prior policy views or are being advanced by someone with similar political values. There are thus strong incentives to bolster policy arguments with scientific claims – good ones if you have them, bad ones if you don't. Moreover, the risk of being discredited for advancing weak, biased, or refuted scientific claims is small, due to the looser standards of evidence and argument in policy arenas than in scientific ones. The result of widespread use of these tactics is that many citizens believe major points of climate science are highly contested and uncertain when they are in fact clearly known and strongly agreed. At the same time, this diversion of the policy debate to specious scientific arguments has stifled discussion of economic and political questions, both positive and normative, on which more vigorous public debate should be taking place.

Although these tactics are unlikely to disappear, their influence can be reduced if participants in policy debates evaluate such claims more skeptically. Treating scientific claims skeptically in policy debates is valuable advice – both for claims promoted as "skeptical" and for others – even though skeptical examination by policy actors cannot match the rigor with which claims are scrutinized in scientific settings. Skepticism toward partisan argument is a virtue, and it is ironic that those advancing distorted scientific claims on climate change call themselves "skeptics," since their success depends on policy-makers and citizens not examining their claims too closely.

Practical skepticism in a policy debate means questioning the origin and foundation of the supposedly scientific claim being made. Who or what is the source, and are there good reasons to regard them as expert and impartial? Holding scientific credentials might be a good indication of expertise, depending on the field, but is no guarantee of impartiality. Is the claim based on published, peer-reviewed research? Claims based on peer-reviewed scientific publications are generally more credible – but this does not mean they are always right – while claims advanced exclusively or primarily in self-published outlets (e.g., non-scientific periodicals, publications of advocacy organizations, newspaper opinion pieces, or blogs) – are generally less credible, unless they are reporting results from other, peer-reviewed sources. Has the claim been verified by additional peer-reviewed studies and widely accepted by the relevant scientific group? Are there opposing scientific views? If so, who holds them – how many people, of what level of relevant expertise – and what are the grounds for saying that one view or the other is correct? Parties to a policy debate should ask these questions, just as scientists ask them in evaluating a scientific claim. Claims that one peer-reviewed scientific paper represents settled knowledge or, more extreme, overturns an established understanding by itself, warrant strong skepticism. So too does any source that uses polemical language or makes personal attacks, that states no limits to the certainty or scope of its claims, or that cannot tell what evidence would weaken its claim. Finally, it is especially important to be skeptical of your friends, and of claims that appear to support your political principles or positions. You are most at risk of being misled by deceptive or erroneous scientific claims that are consistent with your own beliefs or policy preferences, whatever these are. But the true state of the world, and the true state of scientific knowledge about it, take no account of political values.

No matter how well policy actors follow this advice, however, scientific claims cannot be evaluated as carefully in policy debates as in scientific settings. This is why the climate policy debate requires authoritative scientific assessments, generated by processes that are credible, legitimate, prominent, and have some protections against partisan attack. As Section 2.5 discussed, assessments summarize, synthesize, and evaluate scientific knowledge to inform decisions or policy debates. For climate change, scientific assessments are provided by the Intergovernmental Panel on Climate Change (IPCC). But the IPCC is coming under increasing attack by many of the same actors who deny the major points of current scientific knowledge about climate change.

In a sense, anyone wishing to maintain a supposed scientific basis for rejecting climate action has to attack the integrity of the IPCC, since not doing so would concede the points of scientific knowledge that underpin current concerns. And the real risk of bias in scientific assessments of environmental issues clearly does warrant examination. Managing an assessment to achieve high scientific standards and also provide useful information for policy is difficult, and assessments have not always succeeded. Moreover, many actors would wish to exercise political influence over IPCC assessments if they could. But the charges against the IPCC have not stood up to scrutiny.

These charges have taken three distinct forms. A few have attacked the IPCC wholesale, claiming it is really not a scientific body but a political one, and is biased to exaggerate risks of climate change and suppress uncertainty and dissent. Even a moment's inspection of the IPCC process and reports shows such broad-brush attacks to be obviously false. The core of the IPCC assessments are comprehensive scientific reviews on specific climate-change topics, each conducted by a team of dozens of experts on that specific topic from multiple nations and reviewed by dozens more. In view of the number, breadth, and stature of the participating scientists, the thoroughness and criticality with which they review the scientific literature, and the rigor with which their reports are in turn peer-reviewed – with all reviews and authors' responses publicly available – the IPCC assessments are the definitive statements of current climate science, widely used as scientific references and accepted as authoritative by virtually all policy actors engaged in the issue.

A second, subtler charge has been that while the underlying IPCC reports are impartial scientific reviews, the summary statements – particularly the Summary for Policymakers, the short, non-technical summary drafted by national representatives in plenary session – misrepresents the full report by exaggerating risks and understating uncertainties and qualifications, injecting an alarmist bias on climate science and an activist policy stance. This charge merits a more serious examination, since there have been past occasions when the summaries of scientific assessments on other environmental issues have misrepresented the main report – although these past occasions have involved summaries understating, not overstating, environmental risks.

But this charge has held up no better than the broader charge of political bias in the entire IPCC. Summaries always suppress detail and qualifications: they have to, because they are summaries. In drafting IPCC summaries, scientific lead authors have veto power over proposed changes, precisely to protect against introduction of error or political bias at this stage. Beyond this protection, the scale of participation and transparency of process ensure that any attempt to bias summary text to favor one group's position would be both offset by opposing pressures, and severely limited by the open nature of the deliberations. Repeated re-examination of IPCC summaries – including a review of the 2001 assessment specifically requested from the National Academy of Sciences by the Bush administration – has found that they fairly represent the full report, given the need to summarize thousand-page reports into a few pages written for a non-scientific audience.

The final charge, leveled against the IPCC and climate scientists more broadly, is that they are exaggerating both confidence and alarm to advance their own interests, mainly to secure more research funding. Initially stated in a best-selling novel – i.e., a work of fiction – this charge has subsequently been presented as fact in multiple advocacy pieces and newspaper opinion columns, most recently after the circulation of a few embarrassing excerpts from an email archive stolen from a British climate-research unit in late 2009. The charge is ludicrous for many reasons, of which the most obvious is that if climate scientists wished to corruptly advance their funding prospects, this would not be an effective way to do it. The major policy implication of current scientific knowledge of climate risks is that serious efforts to limit and respond to climate change are warranted. But what would such efforts mean for scientific research? Efforts would likely include large increases in research on climate-safe energy technologies, and on sectors affected by climate impacts and ways to reduce their vulnerability. But in a budget-constrained world, this is as likely to mean less support for *climate* research as more – even if there is also a good case for expanded research on climate science. Consequently, when climate scientists state that current knowledge supports serious efforts to limit climate-change risks, they are arguing against their own professional interests. This should lend their conclusions greater credibility.

All these attacks on the scientific assessment process fall into one of two errors. The first error is to imagine the world's climate scientists could possibly conspire to lie about what they know – no matter if their motive is to impose their policy preferences on the world, to advance their corrupt interests, or that they are somehow coerced by activist governments – and to keep this secret. Even if they wanted to, there are multiple reasons why they could not. Science thrives on open exchange of ideas and arguments, and rewards those with persuasive new ideas and contrary claims – so long as they play by the rules of fair argument and evidence. The scientists participating in the IPCC are numerous, diverse in their nationalities and political views, and much more knowledgeable and respected than the few eccentrics and opportunists who are making conspiracy charges. The notion that such a group could be tempted, deluded, or intimidated into unthinking support for erroneous scientific claims – and that they could then keep such corrupt behavior secret – is insupportable.

The second basic error is failing to understand the significance of the IPCC being a scientific assessment body, not a purely scientific body. Assessment bodies sit between the worlds of science and policy, and must manage the resultant tensions. The task of synthesizing current knowledge in a form that is faithful to underlying science but also useful for policy is different and more challenging than writing a scientific review. The IPCC has pursued this balance through procedures by which governments are nominally in charge but exercise no influence on the full assessments, which are entirely scientific reviews. For all the challenges they face, IPCC assessments have managed these tensions effectively. Their deliberations have maintained an impressive level of independence from political interference, despite an organizational structure that could have threatened such independence. To the extent that true synthesis statements of the state of scientific knowledge about climate change exist anywhere, it is in the IPCC assessments. They, and other scientific assessments that achieve similar quality of participation, deliberation, and peer review, are the "gold standard" of trustworthiness of policy-relevant scientific statements, and policy actors can do no better than to rely on them.

In fact, if any bias operates in the IPCC process, it likely arises from scientists' general conservatism in evaluating new claims, and thus operates in the opposite direction from what these criticisms have alleged. On the one hand, such conservatism grants a massive, grave authority to the assessments' conclusions. On the other hand, it may account for the one way the IPCC may be falling short of its responsibilities: through an excess of scientific caution, it may not be speaking clearly enough on points of high policy significance that are not fully resolved scientifically. The most serious controversies over the fourth assessment were of this character: omitting the possibility of ice-sheet collapse from projections of sea-level rise, and omitting the highest emission scenarios from summary graphics of future climate-change projections.

Generic uncertainty arguments

Denying specific points of scientific knowledge and attacking the integrity of scientific assessments are both relatively crude and risky tactics. Sophisticated opponents of action have instead sometimes sought to frame their arguments as "scientific" without advancing any particular scientific claim that might be refuted. A powerful way to do this is to refer to uncertainty in climate science in general, rather than resting on any specific point. This has been a common debating tactic on many environmental issues, but it has become especially prominent since the 2004 US elections, when a leaked strategy memo advised Republican candidates how to deal with climate change. The memo provided a strikingly direct statement of this strategy and its objectives.

"The scientific debate remains open. Voters believe that there is no consensus about global warming within the scientific community. Should the public come to believe that the scientific issues are settled, their views about global warming will change accordingly. Therefore, you need to continue to make the lack of scientific certainty a primary issue in the debate ... The scientific debate is closing [against us] but not yet closed. There is still a window of opportunity to challenge the science."[1]

The argument makes three points. First, there are large uncertainties in scientific knowledge of climate change and human influences on it. Second, given such overwhelming uncertainties, it would be irresponsible to undertake costly actions to limit emissions when we do not know if these are warranted. Third, we should instead seek to reduce uncertainties through further research before making irreversible commitments.

The first claim, that there is much uncertainty in present scientific knowledge of climate change, is correct. As President Bush stated when he announced the United States would not ratify the Kyoto Protocol, "(W)e do not know how much effect natural fluctuations in climate may have had on warming. We do not know how much our climate could or will change in the future. We do not know how fast change will occur, or even how some of our actions could impact it ... And finally, no one can say with any certainty what constitutes a dangerous level of warming and therefore what level must be avoided."[2] These statements were all true when he made them in 2001, and remain true today.

But characterizing this uncertainty as overwhelming, or suggesting there is nothing important about climate change that we do know, is simply false. As Chapter 3 discussed, there are many points of climate science on which knowledge is quite advanced, and several key points – such as whether the climate is warming, whether human activities are primarily responsible, and whether the warming is likely to continue – that are essentially established beyond doubt. We do not know everything – which is what it would mean for there not to be scientific uncertainty – but we know a great deal.

Moreover, the key point of this argument – that justifying action requires eliminating scientific uncertainty, or at least achieving some higher level of confidence and precision about future climate and its impacts than present knowledge provides – is not a scientific argument at all, but a normative judgment about when costly efforts are justified to avoid an uncertain risk. The argument presumes that the status quo, no action, should persist until it is demonstrated that taking action is preferable. Moreover, by citing "scientific uncertainty" as the reason for not acting, the argument presumes the required standard of demonstration is total or near total elimination of uncertainty.

[1] See J. Lee, "A call for softer, greener language," *New York Times*, March 2, 2003, p. 1. The complete memo is posted online by the Environmental Working Group, at http://www.ewg.org.

[2] Remarks by President G.W. Bush, White House Rose Garden Briefing, June 11, 2001.

This argument relies on an analogy, sometimes explicitly stated but more often implied, to other decision domains where we require a high standard of evidence to justify a particular decision. The two most prominent of these analogies are to criminal law and scientific research. The rules of criminal trials require that the defendant is presumed innocent unless the prosecution demonstrates guilt "beyond a reasonable doubt." In scientific research, as discussed in Chapter 2, when a new result claims to contradict current knowledge, it is not accepted until demonstrated to a highly persuasive standard and verified by multiple, independent replications.

In both these cases, the requirement for a high standard of evidence is based on a normative judgment of the relative severity of the two possible kinds of error. In any decision under uncertainty there is unavoidable risk of making the wrong choice, but errors can occur in more than one way. A criminal verdict can be wrong by convicting an innocent defendant, or by acquitting a guilty one; scientific judgment can be wrong by accepting a new claim that turns out to be wrong, or refusing to accept one that turns out to be right. Criminal trials require guilt to be shown beyond a reasonable doubt, thereby biasing the decision in favor of the defendant, because society has long judged it worse to convict an innocent defendant than acquit a guilty one. In science, requiring new claims to be strongly verified reflects a similar balancing of the cost of the two possible errors. Accepting an incorrect new claim is highly costly, since it can confuse and misdirect subsequent research, and cast doubt on the accumulated body of prior knowledge. But failing to accept a correct new claim is less costly, because such rejections are always provisional. A correct claim that is not initially accepted usually accumulates further supporting evidence until it meets the standard for acceptance, so the cost of this high standard is only a delay in accepting the claim until more evidence is obtained.

In both these domains, the crucial point is that the decision rules are chosen intentionally, based on normative judgments of which error is worse. The worse we judge a particular error to be, the more we try to make that error unlikely by biasing the decision process against it. In doing so, we willingly accept an increased risk of making the other error, because we judge it to be less bad.

In other decision areas we use different biases, reflecting different judgments of how bad it is to err in each direction. In civil law – private lawsuits in which only monetary damages or restrictions are at stake, not life or liberty – there is no clear basis to judge it worse to err one way or the other (favoring the plaintiff or the defendant), so civil suits are decided without bias, on the basis of "the preponderance of the evidence." National security policy in the US and other nations has often supported extremely high-cost actions to defend against even unlikely threats, because the cost of being unprepared to meet a threat that does materialize is judged to be so severe.

Decision procedures can be biased either for or against action in any policy area: environmental activists sometimes make the same arguments for a pro-action bias as is commonly advanced for national security. This is not a scientific choice, however, but a judgment of what errors we care more about avoiding. The argument against climate action based on general uncertainty proposes an extreme bias against action: do nothing until we know it is necessary to avoid severe climate-change impacts. Clearly, this approach carries a high risk of not acting enough, or soon enough, because the level of confidence demanded to justify action might not be attained until severe impacts have already occurred or are unavoidable. This could still be the right course, but only if we judge it much worse to act too strongly than not strongly enough – i.e., that the costs of too much mitigation are much worse than the impacts of too much climate change.

But as Chapter 4 summarized, there is no basis for believing this to be correct: in fact, the reverse situation appears more likely. To illustrate this, consider the two extremes. First, suppose climate change and impacts turn out to lie at or below the bottom of the present uncertainty range: in other words, suppose we are as lucky about climate change as we plausibly could be. In this case, an aggressive mitigation program would impose unnecessary costs, probably between a few tenths of a percent and several percent loss of future GDP, depending on how aggressive the program is – although since costs would be spread over a century, even in this case much of the cost could be avoided by scaling back future efforts once we learn they are unnecessary. But in the opposite extreme – i.e., if we are unlucky and climate change and impacts lie at or above the top of the present uncertainty range – then not pursuing aggressive mitigation would impose costs and risks much more severe than these relatively modest costs of over-controlling, including a growing possibility of abrupt or catastrophic changes.

With high and uncertain stakes on both sides, a prudent climate response requires treating costs and risks symmetrically, balancing the risks of doing too much and too little. In this, climate change resembles other high-stakes policy issues requiring action under uncertainty, including responding to security threats such as hostile foreign powers or terrorism, making economic policy, and managing all kinds of risks to life, health, and safety. In climate, as in all these areas, the mere presence of uncertainty does not mean the right response is to do nothing until decisive evidence compels it, any more than the first hint of a remote risk means the right response is maximum feasible efforts, regardless of cost. Rather, the risks, costs, and benefits on both sides – those from doing too much, too soon, and those from doing too little, too late – must be compared and balanced. In assessing this balance for climate change, all serious analyses have found early mitigation action to be warranted: they differ only in how much action they find optimal how soon, and even in this their disagreement is surprisingly small.

5.3 So what should be done? Recommendations for an effective response

In our view, present knowledge and evidence of the risks of climate change are sufficient to justify strong action, starting immediately, despite the uncertainties that remain in all aspects of the climate issue. Given the risk of serious, slow-to-reverse harms, it would be irresponsible to await precise knowledge of the form and magnitude of climate-change risks before taking action to forestall them. The arguments now being advanced against significant near-term action, as summarized above, are either wrong, misleading, or – to the extent they are correct – do not make the case for further delay.

But what action? Even accepting the need to act, there is substantial uncertainty and disagreement over what specifically should be done. As Chapter 4 discussed, designing policies that effectively limit climate change, limit costs, and are politically feasible and sustainable, requires multiple choices that are contested and uncertain. How should effort be allocated between policies that target emissions directly, and those that promote innovations to make future emission cuts easier? For policies that target emissions directly, what is the best choice among conventional regulations, emissions taxes, cap-and-trade systems, or combinations of these – in single jurisdictions and across borders? How can confidence be built in implementation of policies and commitments, particularly between multiple countries with different capacities and cultures? What institutions and processes are needed to evaluate and adapt policies over time, in response to new knowledge and changed conditions and capabilities? And finally, how should the required efforts and burdens be shared, among nations and over time? These matters require decisions, explicitly or implicitly, to move past the present deadlock.

We do not claim authoritative answers or magic solutions to these problems. But present knowledge provides a reasonable basis to identify some choices as more promising than others. In this final section, we present a set of recommendations for actions that in our judgment hold the most promise. Although adaptation and other elements will be essential components of an integrated climate response, our recommendations focus on mitigation initiatives – in line with the current policy debate, and our judgment of where the most urgent action priorities lie.

A complete mitigation strategy would include four elements: goals to define the scale of the required medium to long-term changes; near-term actions that take initial steps toward these goals; a political strategy to build needed participation over time, from first steps to completion; and processes to evaluate and adapt goals and actions in light of changing knowledge, experience, and capabilities. These elements are not completely separate in practice, nor need they be decided in any particular order, but they provide a useful categorization. We present our recommendations for each element in turn.

5.3.1 Medium and long-term goals

Limiting climate change requires transforming the global energy system to rely on climate-safe energy technologies. Such an enormous, multi-decade endeavor is hard to sustain or even understand without stating a goal. Recognizing this, negotiators of the Framework Convention included an explicit goal in Article 2: stabilizing greenhouse-gas concentration "at a level that would prevent dangerous anthropogenic interference with the climate system." This is a fine goal, which gains nearly universal agreement when stated at this level of abstraction. But there is no bright line that demarcates dangerous interference with the climate system, due to both scientific uncertainty about the impacts of different concentrations, and normative disagreement over what impacts are acceptable and what efforts are worth making to avoid them. Partly due to these conceptual difficulties and partly due to political conflict, subsequent negotiations did not succeed at making this objective more operational until the 2 °C target in the Copenhagen Accord, the effect of which remains uncertain.

Outside international negotiations, however, there has been substantial progress in building support for reasonably prudent climate-limitation goals. As Chapter 4 discussed, proposed goals have included limiting warming to 2–3 °C above pre-industrial temperatures, or associated limits on radiative forcing (e.g., 2.5–4.5 watts per square meter) or greenhouse-gas concentrations (e.g., 450–550 ppm of CO_2-equivalent), with some recent suggestions that targets should be lower still, e.g., 1.5 °C, 2.3 watts or 350 ppm. Such targets in turn imply a range of emission trajectories, with world emissions peaking within a decade or two then declining through the rest of the century. For example, the scenarios presented by the IPCC showed stabilizing at 450–500 ppm CO_2-equivalent required emissions to drop 50 to 80 percent from 2000 levels by 2050. Whether expressed as warming, forcing, or concentration, the strictest of these targets carry lower climate-change risks but require sharper reductions of emissions sooner – meaning greater costs and efforts. In our judgment, the low end of these ranges, around 450 ppm CO_2-equivalent, appears to reflect the most prudent balance of uncertain mitigation costs and climate impacts. Such an admittedly ambitious goal would also retain the option of shifting to even stricter limits on climate change, in case unlucky possibilities high in the distributions of climate sensitivity, long-term feedbacks, or impacts should be realized.

Similar goals have been widely proposed by independent analysts and organizations, and increasingly endorsed by national political actors in multiple countries. These goals are all expressed in global terms, however, so deriving goals for particular countries or regions requires additional assumptions about the distribution of effort. For example, the target of cutting US emissions about 80 percent from 1990 levels by 2050 – similar to goals stated by the Obama campaign and current legislative proposals – is consistent with stabilizing concentrations at 450 ppm of CO_2-equivalent, under the assumptions of proportional burden-sharing among industrialized countries and gradual adoption of mitigation commitments by developing countries over the next 20 years.

Setting explicit goals can provide several benefits to a climate strategy. Challenging but attainable goals can motivate action, focus attention, and provide a context for evaluating and choosing near-term measures, even if the relationship between near-term measures and long-term goals is uncertain. Explicit goals can also provide a basis to hold decision-makers accountable. But discussions of climate goals have been complicated by failure to recognize important distinctions that arise when a goal's achievability and the effort required to meet it are uncertain. Challenging but attainable goals can be powerful motivators, but if they really are challenging they will often not be met. Such goals are often used in other settings, e.g., athletics or technology development, but are rare in policy or political settings, where goals are typically stated to hold actors accountable. Goals for accountability do not state aspirations, but baseline expectations: they are expected to be met, and unfavorable consequences follow if they are not – if not material sanctions, then censure or embarrassment, which governments also seek to avoid. Under uncertainty, even actors committed to limiting emissions may reject ambitious goals if these are understood as expectations rather than aspirations. Consequently, if goals are used for accountability or there is a risk they may be (e.g., by activists seeking to retrospectively redefine aspirations as requirements), actors are likely to accept only weak goals they are highly confident they can achieve.

With goal statements subject to such maneuvering, it has been difficult for political bodies to use them effectively. In our view, the benefit of stating explicit goals outweighs the risks. National governments and other decision-making bodies should state goals for mitigation, preferably as default emission trajectories that integrate near, medium, and long-term reductions. Although there must also be procedures to change the trajectory as new knowledge or changed conditions warrant, merely stating a default will shape the expectations of investors and others making long-term decisions, which is crucial to achieving the required changes.

Several conditions must be met, however, to exploit the motivating power of goals without asking governments to take on unacceptable risks. Goal statements should be explicit about specific policies and actions proposed and their assumed effects on emissions, including uncertainty. Goal statements should also state explicit assumptions about other conditions influencing the effectiveness of policies and actions, including mitigation efforts by others. This requirement would be similar to a requirement to state the economic assumptions underlying long-range government budget projections. It may also be useful to explicitly declare two goals that differ in ambition and in expectation of achievement – an expectation and an aspiration.

But while the value of goals as signaling and motivational devices at the national level is clear, their value in international negotiations is less so. Here there is more risk that negotiations over goals will become a contentious distraction from more urgent decisions on near-term measures, particularly since goals will have to be adjusted over time. Perhaps the right compromise is to negotiate goals that are stated with some vagueness, so they reflect a shared but approximate understanding that emissions must be reduced *a lot* by the middle of the century. Nor is it necessary that all national goals be uniform. Rather, we expect them to provide benchmarks for an ongoing debate, in which the most activist governments state strong goals, others weaker ones. If informal coalitions coordinate on common goals, measures to pursue them, and supporting analysis, fine – but not at the risk of delaying concrete first steps. Waiting for full agreement on goals, like waiting for elimination of scientific uncertainty, would mean waiting so long that many desirable goals become unattainable. While we support continuing dialogue on goals, this must occur in parallel with discussions and decisions on concrete actions, which proceeds without waiting for full agreement on goals.

5.3.2 Near-term actions

Whatever goals are adopted, no progress will be made toward the required emission reductions without someone taking the first steps. What should these be, and who should take them? In this, we propose a substantial departure from recent practice. Since climate policy debates began in the early 1990s, it has been widely assumed that the primary location for action would be international negotiations, which would motivate, enable, and coordinate national policies. This approach reflected the recognition that the problem requires a global solution, and the concern that if nations acted separately, those that do the most would see their efforts thwarted by loss of competitiveness and emissions leakage.

But whether or not these concerns were justified – and recent studies suggest they have been over stated – this approach has achieved virtually no progress in reducing emissions. In our judgment, increased leadership through national actions holds greater promise for breaking the current deadlock. Effective mitigation policies must be globally coordinated to be effective – eventually – but full global participation and coordination are not needed at the outset, and waiting to achieve them may serve only to delay the required first steps.

We consequently propose that the first priority should be announcement of a coherent, effective mitigation strategy by the United States, matching or surpassing the climate leadership thus far exercised by the EU, to jump-start the pursuit of serious, broad-based international agreements. Initiatives by other major industrial economies to announce similar strategies in parallel would be welcome, but the initial development and announcement of a US strategy should not await coordination with others.

This mitigation strategy should follow the principles and structure laid out in Chapter 4. It should promote private investment and technology development in climate-safe energy resources by creating clear, consistent incentives. It should take a long-term focus, starting immediately but gently, and increasing the stringency of policies over time, to provide the required sustained, transparent incentives for long-term investments and innovations, to limit costs, and to provide stability for investment planning. And it should seek to minimize costs by allowing flexibility in private actors' implementation, harnessing market forces to the maximum extent consistent with an effective policy. Such an initiative would face fair political challenges, but would be consistent with the declared intentions of both the Obama administration and key Congressional leaders. In its specifics, the initiative we propose is slightly stronger than the bills moving though Congress in 2010.

The mechanics of this national mitigation strategy should include the three elements discussed in Chapter 4: economy-wide, market-based policies to put a price on emissions; support for climate-safe energy research and development; and other regulations and measures as appropriate – plus institutions and mechanisms to review and adapt these measures over time in light of new knowledge, experience, and capabilities. The economy-wide, market-based policies are the central element. These do the main work of creating incentives for investment and technology development, by signaling that greenhouse-gas emissions will grow increasingly costly over the life of current investments. They should be deployed as broadly, consistently, and transparently across the economy as possible. They should start gradually to avoid disruption, but rise consistently following a clearly announced schedule of increasing stringency extending out at least several decades.

The two main options for a policy to create these emission prices are a carbon tax, which would set the emissions price explicitly, or a cap-and-trade system, under which the emissions price would be set by trading of permits. As Chapter 4 discussed, taxes are now widely believed to work better and carry less risk, but nearly all political momentum, in the US and elsewhere, favors cap-and-trade systems. If there were the opportunity to choose between these, one would have to balance the recognized risks of cap-and-trade systems against the greater political obstacles to carbon taxes, asking whether a carbon tax could be made politically feasible and how much it would have to be impaired to make it so. Given the present momentum, however, the preferred course appears to be to design a cap-and-trade system to be maximally effective, efficient, and transparent, and avoid identified risks as much as possible – and to monitor its implementation and effectiveness carefully, ready to change course if necessary.

For the US, a suitable cap might start with reductions of 5 to 10 percent from the projected emissions trajectory within three to five years, with a pre-announced schedule of reductions toward a strict future goal, e.g., 80 percent cuts in 2050. The equivalent emissions price might start around $35 to $70 per ton carbon-equivalent ($10–$20/ton CO_2) and increase by 5 to 10 percent annually in real terms. Permits need not specify emissions in a single year, but can instead allow one ton to be emitted at any time over an extended, even unlimited, period. Equivalently, if permits are of shorter duration, permit holders should be allowed to bank them freely for future use, and to borrow against future emissions subject to adequate provisions to limit fraud and risk of default.

To ensure that a cap-and-trade system provides the required long-term incentives, it should include explicit limits to the price range at which permits can be traded – a price ceiling and floor – which both increase over time according to a pre-announced schedule as the emissions cap tightens. The price ceiling is the "safety valve" – a price at which the government will sell extra permits in any amount. This limits the financial risk of price spikes if the cap is tightened too fast, relative to the pace of development and production of climate-safe technologies. Many emitters have demanded a safety valve, while many environmental groups have opposed it because they want the certainty in total emissions that a firm cap appears to provide. In our view, it is important to include a safety valve to limit disruption from extreme price spikes. (And in any case, the seeming certainty of an emissions cap is illusory: if prices spike high enough, the cap will be loosened or suspended, even if there is no explicit safety valve.) But the level of the safety valve must be high, starting at double to triple the expected initial price level, so it is only reached under conditions so extreme (and unlikely) that temporarily loosening the cap truly is a reasonable short-term response.

The price floor would be a pre-announced lower price at which the government would buy back and retire unlimited quantities of permits. The price floor limits the risk that the cap is set too loose and so fails to provide the required incentives. An initial level for the price floor might be half the expected price, perhaps $15–20/tC. To create the required long-term incentives for investment and technology development, it is crucial that the price ceiling and floor both rise over time according to a pre-announced long-term trajectory, increasing by perhaps 5 to 10 percent per year, as the emissions cap is tightened. If they do not, then as the cap is tightened the system would simply come to operate as a carbon tax at the safety-valve price.

A second requirement to promote maximal effectiveness of a cap-and-trade system is to distribute as many of the permits as possible by auction, rather than giving them free to current emitters. While theory says that this makes no difference to emitters' marginal incentives to emit – the opportunity to sell a permit received for free gives as strong an incentive to cut emissions as the need to buy a permit in order to emit – these situations may differ strongly under reasonable assumptions of how emitters' actual behavior might depart from theory. Auctioning permits is also necessary to avoid the large transfers of wealth to emitters that occur when permits are distributed free. If auctioning all permits initially meets too much resistance, then some fraction should be auctioned, with a firm schedule to move to full auctioning over time, within no more than ten years. This commitment should be made as binding as possible, e.g., by making government budget planning depend on presumed revenues from the auctions.

Beyond a cap-and-trade system or carbon tax, other measures are needed to make a complete strategy: public support for research, development, and demonstration projects (RD&D) in climate-safe energy technologies, and additional regulatory measures. Research support is necessary because even with a price on emissions, research and development in climate-safe technologies produce spillover benefits that are not captured in private incentives. Contrary to some proposals, however, a climate strategy cannot rely exclusively on research and technology development, without the economy-wide measures to put a price on emissions. The reason is that even an aggressive research program is likely to leave climate-safe technologies more costly than conventional alternatives, so they will not be deployed as needed without policy-generated incentives or requirements to do so.

The additional regulatory measures – often called "sectoral measures," because they target specific, high-priority industry sectors or technologies – will enhance the effectiveness of mitigation measures in areas with high technical potential but where energy-market incentives have limited effectiveness. As Chapter 4 discussed, examples include building codes, vehicle efficiency standards, and zoning and planning decisions. To avoid these imposing inequitable burdens on particular sectors, the stringency of sectoral measures should be coordinated with market-wide incentives through life-cycle cost assessment or related analytical techniques. Use of sectoral measures may offer advantages in international negotiations, by allowing coordination among trading partners of the regulatory burdens imposed on particular traded sectors.

5.3.3 Strategic sequencing: toward a global greenhouse-gas strategy

Such an initiative, by the United States and possibly others, would immediately give new energy to international negotiations, but America alone cannot solve the climate-change problem. Any such US initiative can only be the first in a strategic sequence of steps leading to the desired endpoint: global participation in a coordinated, effective, and cost-effective climate-change regime. To reach this goal, each step must facilitate subsequent steps, by motivating the required private investments and providing adequate incentives for current participants to meet their commitments and for additional participants to join.

The second step in this sequence – taken immediately after announcement of the unilateral initiatives we propose, or even earlier in parallel with their development – is for those making these initial commitments to begin negotiations with other nations to extend, strengthen, and coordinate their mitigation commitments. But with which other nations, and in what forum? At this point, the most basic choice is whether to conduct these negotiations within the existing Framework Convention and Kyoto Protocol process. Here, the advantages of working within an existing institutional structure, including two existing treaties that formalize several modes of flexible international action, must be weighed against the disadvantages that have obstructed progress so far – including a history of polarized debates, well entrenched opposition within the US, a record of hasty decisions made without assessment of feasibility or cost, and the loss of credibility that follows from so many nations failing to meet their commitments. The lack of progress after Bali on long-standing points of conflict, particularly the relationship between industrialized-country and developing-country commitments, as well as the procedural deadlocks and rejection of the Copenhagen Accord in the Copenhagen meeting itself, suggest that even adoption of the Accord by major emitters may not open up much room for progress in this forum.

Moreover, any attempt to pursue a serious mitigation agreement through this process must deal with its most basic structural element – universal participation. Both the Framework Convention and Kyoto Protocol have sought the broadest possible participation. The Convention now has more than 190 parties, the Protocol more than 180. Universal participation enhances the legitimacy of the process, and broadens opportunities to cut where it is cheapest, thereby lowering the cost of achieving any specified target. But universal participation, together with procedures or norms of decision-making by consensus, also creates powerful opportunities for obstruction. As Copenhagen re-confirmed, in a universal forum, a group willing to cut emissions cannot negotiate how to do so without others who are not participating having a voice, or even a veto.

This mismatch between who accepts commitments and who has a say in their terms obstructs effective decision-making in multiple ways. It can separate negotiations from considerations of practicality, since many parties are negotiating terms they will not have to live with. It can derail negotiation of initial commitments through other parties' efforts to secure favorable precedents or otherwise maneuver for long-term advantage. It can obstruct attempts to negotiate expansion of the group accepting commitments, or to develop incentives to motivate such expansion. Most seriously, universality empowers some states, such as major fossil-fuel exporters, who may oppose the entire endeavor. These may have an interest in obstructing negotiations, as they prefer not just to avoid cutting their own emissions, but also to prevent others from doing so. In a universal forum these nations cannot be excluded from mitigation negotiations, while the norm of consensus decision-making means that their obstructive tactics are frequently effective.

Because these unavoidable disadvantages of a universal forum are likely to continue hindering effective action, we hold little hope for serious progress in negotiations under the existing Framework Convention and Kyoto Protocol, or any other forum with universal participation. Discussions here should of course continue, particularly as strong actions by major emitters in support of the Copenhagen Accord may galvanize more progress in 2010 under the Convention and Protocol than we expect. But we doubt that the required rate of progress can be achieved here, and so recommend not relying on this as the only forum for climate negotiations. Rather, serious near-term action should be pursued through some smaller forum, in which there are no bystanders: those participating and shaping actions are the same as those undertaking actions.

What nations should be involved in this smaller forum? The participating group must be big enough to make a substantial contribution to global mitigation and representative enough that a deal among them provides a plausible model for a subsequent global agreement. Yet the group must be sufficiently few in number to allow effective negotiations with real exchange of benefits and obligations while limiting opportunities for posturing and obstruction. While various proposals have been advanced, we believe the most promising approach would be negotiations among a group of the world's largest emitters and economies among both the industrialized and developing countries, numbering between about a dozen and twenty nations. Similar in makeup to the G-20, and to the groups convened for informal climate consultations by both the Bush and Obama administrations, such a group could represent about two-thirds of current global emissions and so could make a major contribution to reducing global emissions. Their economies would be large enough to create powerful incentives for firms and investors to develop climate-safe technologies, and to limit risks of emissions leakage. And they would be diverse enough that a climate-change agreement acceptable to them would provide a plausible model for a subsequent global agreement. Yet they would be few enough to allow the possibility of a serious negotiation, with all participants expected to share in responsibilities as well as benefits, and without the opportunities for obstruction provided by larger, more formal proceedings.

This group would negotiate a set of linked actions on climate change and related issues that are acceptable to all participants, and which outline the shape of a subsequent global climate deal. The main task would be to link internationally the three key elements of national mitigation strategies: economy-wide mechanisms to put a price on emissions, support for research and development, and sectoral regulations. As at the national level, the economy-wide mechanisms are the primary element, and extending these internationally would be the most important task in building an effective multilateral mitigation effort among this group.

These negotiations of mutual mitigation effort must also address the conflict between industrialized and developing countries at the heart of the current deadlock. Effective mitigation requires global efforts. But nearly all developing countries continue to reject national emission limits (although making many sectoral efforts), while industrialized countries are unwilling to take serious steps without assurances that developing countries will share the burden. In our view, the most promising approach to resolve this deadlock would be simultaneous, linked negotiations of mitigation commitments for all participants, but with the strongest commitments for developing countries coming into effect only after a delay and with specified conditions. Under this approach, industrialized-country participants would accept firm mitigation commitments which come into effect within a few years, perhaps on the same schedule as the unilateral initiative proposed for the US above. Developing-country commitments would be negotiated and adopted at the same time, but would become blinding later, and only if specified conditions are met. The conditions would include prior mitigation by industrialized countries, and possibly others such as measures of development progress or indicators of the severity of climate-change risk. By ensuring that developing-country burdens reflect their different status and do not obstruct their development, this approach would be consistent with the Framework Convention's principle of common but differentiated responsibility.

Although this approach begins with asymmetric efforts, its aim is to move toward a situation in which obligations are symmetric: all contribute in expectation that all others contribute, and each participant's obligations depend on others meeting theirs, with mutual review, assessment, and policy analysis to help all parties meet their goals effectively and at minimum cost. The approach does impose some risk on the industrialized-country participants, who must make unconditional mitigation efforts immediately, while the reciprocal efforts from the developing countries are delayed and contingent. But the specific terms of negotiations could limit this risk in various ways, by making some developing-country commitments unconditional (e.g., sectoral mitigation policies, which in many cases would merely extend initiatives already underway), or by initially holding industrial countries accountable only for the policies they promised, not necessarily the resultant emission cuts expected. In addition, the risk to industrialized-country participants would be limited by the fact that they will not continue their efforts if others do not meet their commitments.

How would these negotiated mitigation obligations be implemented in practice? In part, this would depend on what combination of cap-and-trade systems and carbon taxes parties have adopted domestically. These negotiations could link cap-and-trade systems by international exchange of permits, with appropriate oversight of permit allocation, trading, and compliance. Such an integrated tradable-permit system would require explicit negotiation of national emission baselines from which rights to trade are defined.[3] Alternatively, these negotiations could coordinate carbon tax levels, scope, exemptions, and interactions with existing tax systems, presumably with tax revenues retained nationally. In either form, specific terms can be negotiated so developing-country participants' overall mitigation burdens phase in over 10 to 20 years as discussed above, and they bear a small share of the cost or even receive a net benefit prior to that point. Whatever the negotiated distribution of burdens, international flexibility mechanisms would allow immediate mitigation investments in developing countries, with clear accounting for their emission effects once national baselines were established. Since different countries may choose different combinations of permit and tax systems, negotiations should also build analytic capacity

[3] Contrary to some proposals, we reject the possibility of meaningful progress in limiting emissions by constructing international permit markets without agreement on permit allocations or emissions baselines. Such an approach would lead to unrestricted allocations and permit prices so low they would fail to provide adequate incentives to reduce emissions, or even to build and monitor the integrity of the trading system.

to assess and compare overall stringency of mitigation effort, and the administrative capacity to integrate different systems.

Negotiations would also include the two other elements of mitigation strategy, research and development support and sectoral regulations. These elements may be even more essential in international negotiations than in domestic mitigation policy, because they provide additional means to limit risks and distribute benefits to construct a deal acceptable to all parties. Negotiated sectoral measures might include common performance standards or emission intensity targets for particular industry sectors or products, e.g., steel or automobiles. These measures can draw on the demonstrated willingness of major developing countries to adopt specific emission-limiting regulations more readily than economy-wide constraints. They can also limit competitive losses or trade distortions in sectors that are emissions-intensive and traded. Negotiations of climate-safe energy research can facilitate emission reductions by promoting development and diffusion of climate-safe energy innovations, while accompanying negotiations over sharing support for such efforts and distribution of resultant property and benefits can provide additional tools to distribute benefits among participants and so build a viable package deal.

These negotiations would be broader than mitigation policy. Although these negotiations would not include the most vulnerable nations, they would still have to address climate impacts and vulnerabilities, through agreements on joint research, assessment, building adaptation capacity, and response and compensation. They may also begin developing a collaborative framework for research, assessment, and regulation of geoengineering. And they may also need to include issues outside climate policy, addressing related aspects of energy policy, technology, finance, trade and investment, and development. The breadth of the agenda should be guided by the need to address the key priorities of all participants and provide sufficient benefits in tandem with obligations to be acceptable to all.

If these negotiations succeed, they would represent a large step toward an effective global response to climate change. But to get there, the agreement reached in this initial negotiating group must facilitate, not obstruct, subsequent expansion toward near-global participation through other nations joining. Moreover, to limit the risks of a transitional period with partial participation, this expansion must occur relatively quickly so the initial group's deal must create appropriate incentives to promote it. To limit the risk that the initial group suffers sustained competitive disadvantage, with resultant shifts of emissions-intensive investment to uncontrolled regions and emissions leakage, the initial agreement must limit incentives for high-emitting industries to move outside the agreement and create incentives for additional countries to join. One way to create these incentives is through trade measures that equalize the mitigation cost burden between goods produced inside and outside the participating group.

The specific form of such measures depends on the mitigation policies adopted. Emission taxes can be adjusted at the border of the participating group, charged on imports and rebated to exports in proportion to the emissions generated in their production. In a cap-and-trade system, the equivalent effect can be achieved by requiring imports to acquire emission permits and granting new permits to exports. The effect of such measures is to equalize the cost of mitigation policy between equivalent goods produced inside and outside the control area. By reducing competitive pressures to relocate high-emitting industry outside, such measures eliminate one major pathway for emissions leakage. By reducing competitive advantages to enterprises outside the control area, they would also reduce the incentives for governments to stay outside.

Such measures would face several challenges. They would have to be judged acceptable under the rules of the World Trade Organization (WTO), and would have to demonstrate sufficient accuracy in attributing emissions to products. These calculations would probably have to be done by an impartial expert body collectively accountable to all participating nations, with procedures for review and appeal of its calculations, thereby limiting the risk of the measures being used as concealed trade protection. They might also have to be flexible enough to take account of trading partners' development status. These are serious challenges, but in our view they can be overcome through careful design of the measures and processes to implement them. Perhaps the best argument for these measures is that if the threat of them being enacted motivates enough other nations to join, the measures would not have to be implemented. Similar use of trade measures was highly effective in motivating nations to join the Montreal Protocol, including one restriction so broad and burdensome that it was eventually deemed impractical. Greenhouse-gas-related trade measures would have a similar aim – to motivate such rapid expansion of participation while they are being studied and developed that implementing them becomes unnecessary.

In sum, the strategic sequence we propose to move from the present deadlock toward global climate action would proceed in three linked phases. First, the United States, perhaps with other major industrial economies, would unilaterally announce a serious set of mitigation commitments. Second, a group of major world economies would negotiate a package deal on climate and energy policy, including mutually acceptable mitigation commitments both sectoral and economy-wide. Third, the structure of this deal would organize subsequent development of a global climate-change agreement, probably by negotiation of an amended Framework Convention on Climate Change or a new Protocol under that Convention. The key to the proposal is not relying on negotiations in a universal forum for initial steps, but rather on a combination of leading unilateral announcements and negotiations in a manageable small group of major world economies.

We do not claim this approach is sure to succeed, nor that the pursuit of stronger agreements under the Convention is sure to fail. But prospects for the current process are unpromising, and the urgency for more serious action is great. A clear lesson of Copenhagen is that leadership by major emitters within the universal negotiations is not enough, because too many opportunities for obstruction exist in the larger forum. Rather, leaders' commitments must be fully negotiated and moving to implementation before returning to the universal forum. Although the negotiating group we propose is similar to those convened informally by the Bush and Obama administrations, achieving a deal of adequate scope and seriousness will require stronger institutionalization and support. One promising institutional model would involve regularizing the L-20 summit process, at the level of heads of government, with continuing national (and perhaps international) staff support. Despite the difficulties, we judge such an approach to have the greatest chance of success. It would allow orderly negotiation of an initial climate-change agreement, under the control of those making commitments. It could provide a feasible path for the required expansion to global participation, while limiting risks to the early movers. And it could provide a continuing institutional base that would be available to address other global problems not adequately managed in existing institutions.

5.3.4 Adjusting responses over time

The near-term US initiative and negotiations we propose may appear highly ambitious, but they are still only early steps toward effective management of climate change, for two reasons. First, the required transformation is of such a scale that it cannot be achieved by near-term actions alone, however bold, but will take sustained efforts for many decades. Second, these near-term steps must be taken under substantial uncertainty, about climate change and its impacts, and about costs and opportunities for mitigation. They consequently cannot lock in a complete future path through the required energy transformation, but will have to be adjusted as we proceed, gather experience, and learn more. The required expansion from the initial negotiating group toward global participation is only one dimension in which initial decisions must adapt over time, probably the simplest and most foreseeable one. In addition, the mix of technologies being developed and adopted, the form and stringency of policies to motivate them, and even the climate-stabilization goal, must all be periodically reassessed and potentially revised in light of experience and advancing knowledge and capabilities. Evidence of higher climate sensitivity, faster climate change, more severe impacts, or lower mitigation costs will call for strengthening mitigation through higher emission prices or tighter caps and regulations, despite the long lags between such efforts and their climatic effects. Contrary evidence – lower sensitivity, slower changes, milder impacts, or higher mitigation costs – will suggest relaxing efforts.

The need to act now under substantial uncertainty creates what may look like a paradox. On the one hand, a long-term default trajectory for future emissions and policies must be stated now, to give investors the required long-term incentives. On the other hand, future advances in knowledge or capabilities may require changing these trajectories as we proceed. The way to resolve this tension is to adjust trajectories but do it carefully, balancing the benefits of adjusting to changed knowledge – i.e., the cost of maintaining an emissions or policy path that has become sub-optimal in view of new knowledge – against the costs of disruption, investor uncertainty, and loss of credibility of announced trajectories caused by changing them. There are various practical ways that policy adjustments can limit this disruption. Changes to the previously announced trajectory might require several years' advance notice, or be limited to some specified percentage change per year. Alternatively, adjustments might treat new and existing capital separately, imposing changes immediately on new investments but granting a grace period for existing capital to come under the new requirements gradually.[4]

Present decisions can anticipate this need for future adjustment, by building flexibility into the structure of near-term policies and agreements. Present decisions can also support research and assessment likely to be useful in informing future decisions, as well as development of technologies likely to strengthen future capacity to manage the issue, e.g., technologies to monitor emissions and activities, such as those related to land-use, that are impractical to control now but must be included as emission limits tighten.

Beyond these obvious ways that current decisions can support future adjustments, the process of reviewing and adjusting policies will also require some set of institutions and procedures to inform and guide future decisions. How should future adjustments be decided, and who should make the decisions? Although the feasibility and legitimacy of today's decision-makers imposing strong constraints on future decisions are disputed, a limited form of such constraint is widely accepted and practiced in other environmental regimes. Such a process, based on controlling future decision-makers' agenda and giving them expert-based information, has been used effectively in the Montreal Protocol, and the outline for

[4] Obviously if this approach is taken, any preference for old capital must phase out over a firmly announced period, to limit incentives to keep old capital in service longer.

a similar process exists in the Framework Convention on Climate Change. This process includes a legal requirement to periodically review policies and evaluate their adequacy relative to evolving knowledge and capabilities. Reports of expert assessment bodies on advancing scientific knowledge and technological feasibility are provided to inform these periodic policy reviews. The Montreal Protocol experience suggests that assessments of technological feasibility may be particularly crucial once policies are in place, as goals shift toward maximal reduction of risk, limited only by considerations of technical feasibility and cost.

5.4 Conclusion

In closing, this book has summarized present scientific knowledge about how and why the climate is changing, how it is likely to change over the coming century, what the associated impacts might be, and what can be done about it. We have documented the extensive scientific evidence that the Earth is warming, that humans are very likely responsible for most of the rapid recent warming, and that climate change will continue with impacts that may be severe within this century. Based on this evidence, it is our judgment that well established risks from climate change call for an urgent, high-priority response, aiming both to reduce future emissions and to prepare for a more uncertain and less benign climate than we have enjoyed for the past century. Concrete efforts to build such a response must begin immediately.

Although there are signs of movement as we write in early 2010, society has not yet mounted a serious response. There are many reasons for this. Some are related to unavoidable scientific uncertainty - which does not justify inaction, but does provide rhetorical opportunities to confuse the issue and advocate delay, and also makes it hard to identify what specific action to take. Some are related to the substantial costs of limiting and managing climate change, and the difficulty of identifying policies that distribute these costs acceptably. Whatever the mix of reasons, present actions are utterly inadequate relative to the gravity of the issue. Although climate change has been known for more than two decades, major nations are only approaching the starting line of taking serious responses. Negotiations and debate at the international level, where the main action must occur, remain largely deadlocked along lines of conflict that have moved only a little, and only recently. Despite the modest signs of hope to be found in the Copenhagen Accord, there is no basis yet for confidence that a breakthrough is near.

In several important respects, our proposed approach to managing climate change follows an emerging consensus. The most urgent priority is enactment of policies by major nations, coordinated and linked through international negotiations, to limit greenhouse-gas emissions through market-based measures that put an appropriate price on emissions, supplemented by support for research and development in climate-safe energy technologies and other sectoral regulations. In one important respect, however, we propose a break from current practice and expectations: instead of continuing to seek international climate action through global negotiations under the Framework Convention on Climate Change, we advocate a major unilateral initiative by the United States, followed immediately by negotiation of a package deal of linked climate and energy commitments among a small group of major industrialized and developing nations. This deal would then provide the core of a global climate negotiation.

But whatever the details of the first serious steps toward managing climate change, it is crucial to take them. In view of how long we have waited already, it is far more important to do something serious than to worry about getting the first step precisely right. Indeed, given the remaining uncertainties and political barriers to effective action, the first steps are sure to be imperfect, and will need to be assessed and adapted over time. Consequently, developing effective institutions and processes to support effective research and assessment and adjust policies as experience is gained, capabilities change, and knowledge advances, is far more important for successful management of climate change than the details of initial policies. Developing these adaptation processes will be a novel challenge, and a crucial one.

Managing human disruptions of the Earth's climate is like piloting a supertanker through dangerous waters. Although we cannot be certain, it looks increasingly likely that there are rocks ahead: in fact, we might be pointed right at one. We know what direction we need to steer, but do not know how far we must steer to avoid this rock, whether there are other rocks around, or how hard we can steer without risking damage to the ship. Moreover, a big ship like this one takes miles to change course. Unfortunately, no one is at the wheel right now. The crew is downstairs, arguing about whether there really are rocks ahead, what the precise course is that we must steer to reach our ultimate destination, and whose job it is to steer. While the crew is arguing, the ship is getting closer to the rocks. Somehow, what we need is to get someone upstairs to start steering us away from the rocks – now. Because the steering is so slow, it must start right away. At the same time, we need to learn more about where the rocks are – and also to learn, by starting to steer, about how the ship responds and how hard we can steer it. But neither of these needs to learn more justifies waiting to start the steering: they just mean we must steer carefully, and be vigilant to all we can learn about the ship and the hazards in the waters, while we do it. We can probably still avoid the rocks, but we need to start now.

Further reading for Chapter 5

Joseph E. Aldy, A. J. Krupnick, R. G. Newell, I. W. H. Parry, and W. A. Pizer (2009). *Designing Climate Mitigation Policy*. Discussion paper 08–16 (May). Washington, DC: Resources for the Future. At www.rff.org/RFF/Documents/RFF-DP-08–16.pdf

A thorough review of recent issues in policy design to reduce emissions, including estimates of costs and benefits, the optimal trajectory of emissions prices, policy instrument design, and the relationship between emissions and technology policies.

Joseph E. Aldy and R.N. Stavins, eds. (2009). *Post-Kyoto International Climate Policy: Summary for Policymakers*. Cambridge, UK: Cambridge University Press.

A discussion of alternative approaches to negotiating new climate agreements after 2012, stressing approaches that might be negotiated in Copenhagen as successors to the Kyoto Protocol.

Council on Foreign Relations (2008). *Confronting Climate Change: A Strategy for U.S. Foreign Policy*. Independent Task Force Report No. 61, G.E. Pataki and T.J. Vilsack, chairs. New York: Council on Foreign Relations.

The most recent of many senior panel reports presenting recommendations for US climate-change policy, with particular emphasis on how US actions can achieve leverage to influence global emissions trends. Three brief dissenting statements compactly capture three of the most acute present dimensions of policy disagreement – how much to integrate forest emissions into near-term steps, whether to favor taxes or cap-and-trade systems, and whether international negotiations should initially stress binding treaties under UN auspices, or political agreements adopted in smaller and less formal settings.

Robert Lempert (2009). *Setting Appropriate Goals: A Long-term Climate Decision*. Workshop, Shaping Tomorrow Today: Near-term steps towards long-term goals. Santa Monica: RAND Pardee Center. At www.rand.org/international_programs/pardee

An innovative discussion of alternative long-term strategies to manage climate change, the assumptions that each depends upon, and the role of alternative types of goals in shaping a strategy that is more robust to uncertainties.

Edward A. Parson (2008). *The Long Haul: Managing the Energy Transition to Limit Climate Change*. Workshop synthesis report, August 2008. At www-personal.umich.edu/~parson/website/research.html

The report of a workshop that examined how climate policies might need to be adjusted over time in response to evolution of knowledge, uncertainties, and capabilities, and what near-term decisions regarding policies and institutions might best contribute to that required future adaptation.

Glossary

Adaptation Any response to climate change that adjusts human society to the changed climate in order to reduce the resultant harms and take advantage of any associated benefits. Examples of adaptation measures include building sea walls or dikes to limit risks from higher sea levels or river flooding, or planting drought-resistant crops to deal with drier summers in agricultural regions.

Aerosols Small solid or liquid particles suspended in the atmosphere. There are many different types of aerosols, including dust and soot. Some are natural and some are released or increased by human activities. All have short atmospheric lifetimes. Aerosols exert many direct and indirect effects on climate, some warming and some cooling the surface. There is substantial uncertainty about the present total climate effect of aerosols, but the best estimate is that they exert a net cooling effect that offsets about half the warming effect of greenhouse gases.

Anomalies, temperature The difference between a measured temperature and the temperature at the same location averaged over a reference time period.

Assessment The process of synthesizing, evaluating, and communicating scientific knowledge to inform policy or decisions. Scientific assessments usually involve a committee of relevant experts who review current knowledge and uncertainty on specified policy-relevant questions and produce a report including simplified summaries understandable to non-specialists.

Baseline scenario A scenario of future emissions or climate that assumes no intentional intervention to change future trends, from which alternative mitigation goals or strategies are evaluated.

Blackbody radiation Radiation from an idealized body that absorbs all photons falling on it and emits photons with a distribution of wavelengths determined by its temperature. The name blackbody comes from the fact that at room temperature the photons emitted are not visible to humans – so the body appears black.

Cap-and-trade A system of regulation in which any emissions source must hold a permit for each ton of greenhouse-gases it emits. The government decides the total quantity of emissions to be allowed and distributes that quantity of permits, either by auction or free distribution. Thereafter, emitters may buy and sell permits among themselves. The price of permits, which is the price emitters face for each ton they emit, is set by trading in this permit market, not by the government.

Carbon capture and storage Technologies that allow using the energy content of carbon-based fuels without emitting CO_2 to the atmosphere, by separating or capturing the carbon and storing it in some secure, long-lived reservoir underground or undersea.

Carbon tax A system of regulation in which any emissions source must pay a fee or tax for each ton it emits. A major difference between carbon taxes and cap-and-trade systems is that the price emitters face for each ton they emit is the tax rate, which is set directly by the government.

CFCs *See* chlorofluorocarbons

CH_4 *See* methane

Chlorofluorocarbons A family of synthetic industrial chemicals, derived from methane or ethane by replacement of all hydrogen atoms by chlorine or fluorine. Used as refrigerants, solvents, aerosol spray propellants, and in various other applications, these chemicals were the main contributors to human-caused depletion of the ozone layer. They are now strictly controlled by an international treaty, the Montreal Protocol. They and several related classes of chemicals, including some developed as less ozone-depleting substitutes, are also powerful greenhouse gases.

Climate Atmospheric conditions, including temperature, humidity, precipitation, wind, averaged over time. The climate of a place is distinguished from its weather, the atmospheric conditions at any particular time. The climate of a place is less variable and more predictable than its weather.

Climate sensitivity The change in climate caused by a specified forcing. The most common measure of the sensitivity of the Earth's climate is the change in global-average temperature that occurs in response to an instantaneous doubling of the pre-industrial concentration of CO_2, from 280 ppm to 560 ppm, followed by a period to allow the climate to achieve equilibrium to this higher CO_2 amount. The current estimate of this doubled-CO_2 sensitivity is 2.0°C to 4.5°C.

Climate variability (also internal variability or natural variability) Changes in climate that occur without any external forcing. The best known example of internal climate variability is the El Niño/Southern Oscillation, or ENSO.

CO_2 *See* carbon dioxide

CO_2-equivalent A single measure to account for the total climate effect of CO_2 and other greenhouse gases, defined as the atmospheric concentration of CO_2 that would have the same radiative forcing as a specified mixture of CO_2 and other gases.

Carbon dioxide (CO_2) The major greenhouse gas being directly increased by human activities. Carbon dioxide is released to the atmosphere by combustion of any carbon-containing fuel source, including fossil fuels and biomass. Since the large-scale exploitation of fossil fuels began in the industrial revolution, the atmospheric concentration of CO_2 has increased from about 280 ppm to about 385 ppm in 2009. It is presently increasing by about 2 ppm per year.

Carbon intensity The amount of CO_2 emitted per unit energy extracted or converted. The carbon intensity of a nation's economy measures its total dependence on carbon-emitting fuels. The carbon intensity of most economies has declined over the past century as lower-carbon fuels such as natural gas, nuclear, and renewables have partly replaced higher-carbon sources such as coal and wood, but this trend of decarbonization has recently stagnated.

Command-and-control regulation Environmental regulation that specifies what businesses or other emissions sources must do to reduce emissions, rather than specifying an environmental goal and granting them flexibility in deciding how to meet it.

Copenhagen Accord A political agreement negotiated by 28 nations in December 2009, which stated new goals and commitments for limiting climate change, reducing emissions, verifying commitments, and finance.

Cost-benefit analysis A method to evaluate the total social benefits of proposed public policies, by comparing the benefits of each proposed action to its costs.

Deforestation The process of clearing land through cutting forests. If the cut trees are burned, the carbon they contain is released to the atmosphere. Total emissions from deforestation are estimated at about 1.6 GtC in 2000 or roughly 20 percent of total human emissions.

Discounting A method to convert costs or benefits occurring at different times to a common scale so they can be aggregated and compared. Conventional discounting multiplies future effects by a constant factor per time period, equivalent to compound interest on a savings account. Discount rates used to evaluate public policies and investment projects typically range from about 1 percent to 10 percent per year. How impacts occurring late this century or later are discounted has crucial and controversial effects on the evaluation of climate-change policies.

Downstream regulation Regulation of greenhouse-gas emissions applied at the point of combustion of a fossil fuel, where the emission to the atmosphere actually occurs.

El Niño/Southern Oscillation (ENSO) The best known pattern of internal climate variability, a large-scale reorganization of atmospheric and ocean circulation centered on the tropical Pacific Ocean that occurs every few years. The two phases of the Southern Oscillation, called El Niño and La Niña, originate in changes in the equatorial trade winds blowing across the Pacific and resultant upwelling of cool water off the west coast of South America. Linked changes in temperature and rainfall extend worldwide, with the Earth's average temperature increasing during an El Niño and decreasing during a La Niña.

Emission Permits *See* cap-and-trade

Emissions intensity (or greenhouse-gas intensity) The amount of CO_2 or other greenhouse gases emitted per unit of output produced, in an industrial process or an economy. The emissions intensity of an activity combines the energy intensity of the activity (how much energy is used to produce a ton of steel or a dollar of economic output) with the carbon intensity of the energy source. At present, while high-income industrialized countries have higher emissions per person than middle and lower-income countries, their emissions intensity, measured as tons emitted per dollar of economic output produced, is lower.

Energy intensity The amount of energy consumed per unit output, in an industrial process or an economy.

Equilibrium (or energy balance) A state in which some quantity is maintained unchanging, because flows that would tend to increase or decrease it are in balance. Energy balance for the Earth's climate occurs when energy input from the Sun and energy output through the Earth radiating to space are equal. In this state, the temperature of the Earth will not tend to either warm or cool.

FCCC *See* Framework Convention on Climate Change

Feedback A process by which an initial change to a system causes further changes. If the further changes amplify the initial change, they are called positive feedbacks; if they decrease the initial change, they are called negative feedbacks. The Earth's climate system has many feedback processes that amplify or dampen the initial warming effect from an increase in greenhouse gases. Overall, the Earth's feedbacks are net positive, and they likely more than double the warming due to greenhouse gases alone.

Fossil fuels Carbon-based fuels derived from fossils of ancient living things. The major fossil fuels are coal, petroleum, and natural gas.

Framework Convention on Climate Change (FCCC) The first international treaty on climate change, which was signed in June 1992 and entered into force in 1994. It states a broad structure and principles for international action on climate change, but contains few specific binding requirements. Nearly all nations in the world are parties, including the United States.

GCM *See* general circulation model

GDP *See* gross domestic product

General circulation model (GCM) A mathematical model, run on a computer, which represents known physical processes to simulate the Earth's climate. These models are used to examine causes of past climate variations, and to project future climate changes in response to specified scenarios of greenhouse-gas emissions and other forcings.

Geoengineering Active manipulation of the global climate to offset the effects of increased greenhouse gases in the atmosphere. Examples of potential geoengineering measures include launching reflective aerosols into the stratosphere to reflect solar energy back to space, or positioning sunshades in space to block part of the Sun and shade the Earth.

Gigaton (Gt) One billion metric tons, where a metric ton is equal to 1000 kg or 2200 lbs.

Gigaton of carbon (GtC) A common unit for measuring emissions or atmospheric quantities of CO_2, an amount of carbon dioxide containing 1 billion metric tons of carbon. In this measure, the mass of oxygen in the CO_2 molecule is not counted. Since one atom of carbon weighs 12 units, while one molecule of CO_2 weighs 44 units, one GtC is equal to 3.67 $GtCO_2$.

Gigaton of CO_2 ($GtCO_2$) An alternative convention for measuring emissions or atmospheric quantities of CO_2. An amount of carbon dioxide of mass 1 billion metric tons, counting the complete mass of the CO_2 molecule, including the oxygen. One $GtCO_2$ equals 0.27 GtC.

Greenhouse effect The process by which trace gases in the atmosphere absorb and re-emit infrared radiation, thereby impeding the release of infrared radiation from the Earth's surface to space and warming the surface. The greenhouse effect is a natural process that warms the Earth's surface to its present comfortable state. Human-caused climate change is driven by increases in atmospheric greenhouse gases increasing the strength of the greenhouse effect.

Gross domestic product (GDP) The total value of goods and services produced by an economy. Per capita GDP (GDP divided by the population) is a common measure of a society's affluence or development status.

Infrared Light, or electromagnetic radiation, with wavelengths ranging from about 0.8 to 100 microns – i.e., longer wavelengths than those of visible light, which are about 0.3 to 0.8 microns. The Earth, because of its temperature, emits almost all its radiation in the infrared range.

Integrated Assessment Analyses that consider climate-change impacts, adaptation, and mitigation together in a consistent quantitative framework, to study the costs and benefits of different mitigation and adaptation strategies.

Intergovernmental Panel on Climate Change (IPCC) The international body responsible for conducting scientific assessments of climate change, established in 1988 by the World Meteorological Organization (WMO) and the United Nations Environment Programme (UNEP). The IPCC publishes comprehensive assessment reports on scientific understanding of climate change every five or six years, plus other reports on specific topics.

Kyoto Protocol The second international treaty on climate change, which was signed in Kyoto in 1997 and entered into force in 2005. The Protocol commits industrialized countries, called "annex 1 countries," to reduce emissions of CO_2 and other greenhouse gases below 1990 levels by an average of 5.2 percent, over the period 2008–2012. The only major nation not a party to the Protocol is the United States.

Land-use emissions Emissions that result from human land-use activities or changes in land cover. They include net emissions from cutting, re-planting, and managing forests, as well as changes in the carbon stock in soil, which can be either increased or decreased by agricultural and forest management practices.

Leakage, emissions Increases in emissions that may occur outside a region undertaking emissions controls, which can decrease the effectiveness of emissions controls. Emissions leakage may occur through shifts in global energy markets, or through movement of investment in emissions-intensive industries to regions not controlling emissions.

Marginal cost The additional cost of some small change in an activity. For example, the marginal cost per ton of reducing emissions 100 tons is the additional cost of changing the reduction from 99 tons to 100. The marginal cost of the 100th unit is generally not the same as the average cost of all 100 units. In calculus terms, marginal cost is the partial derivative of total cost with respect to changes in the level of action.

Market-based policy mechanisms Policies that give businesses or other emissions sources an incentive to reduce emissions to achieve some overall environmental goal, but which grant them flexibility in deciding how to meet it. The major examples of market-based mechanisms are emissions taxes and cap-and-trade systems, but other forms are possible.

Methane (CH_4) A greenhouse gas emitted from several natural sources, plus human emissions from rice paddies, landfills, livestock, and the extraction and processing of fossil fuels. While emitted in much smaller quantities than CO_2, methane contributes about 20 times more warming per pound emitted, so its contribution to current warming is about one-fourth that of CO_2.

Metric ton 1000 kg or 2200 lbs, sometimes written "tonne."

Mitigation Activities that aim to slow or stop climate change by reducing the emissions of CO_2 or other greenhouse gases that are responsible.

Nitrous oxide (N_2O) This is an important greenhouse gas, which is emitted from natural as well as various agricultural and industrial processes. While emitted in much smaller quantities than CO_2, it contributes substantially more warming per pound emitted, so it nonetheless plays an important role in the climate change problem.

Normative statement A statement of evaluation, which says that something is good or bad, right or wrong, desirable or undesirable, just or unjust, and so on. Normative statements are distinguished from positive statements.

Ozone layer A region of the atmosphere with elevated concentration of ozone (O_3), lying roughly between altitudes of 15 and 25 kilometers. The ozone layer protects life on Earth's surface by absorbing most of the high-energy ultraviolet (UV) radiation in sunlight.

Parts per million (ppm) A unit for expressing abundance of trace gases in the atmosphere. An abundance of 1 ppm means that there is one molecule of the gas of interest for every one million molecules of air. The present atmospheric abundance of CO_2 is about 385 ppm, meaning that 385 out of every million molecules, or 0.0385 percent of the atmosphere by volume, are CO_2.

Peer review The main process used for evaluating scientific papers before publication, by which experts in the specific subject being discussed, usually anonymous, are asked to critically review submitted papers to look for any errors or weaknesses.

Positive statement A statement about the way things are, rather than what ought to be. Positive statements might concern some state of affairs ("It is raining"), a trend over time ("Winters are getting warmer"), or a causal relationship that explains why something happens ("Smoking causes cancer"). They are distinguished from normative statements.

Proxy climate record A record of past climate that has been imprinted on some long-lived physical, chemical, or biological system. Climate proxies can provide evidence of climate conditions before the modern instrumental record. Major forms of climate proxy records include tree rings, ice cores, corals, ocean sediments, and boreholes.

Radiative forcing A change in the net energy flow reaching the Earth's surface and lower atmosphere, measured in terms of energy flow (watts) per square meter, due to some specified change in the climate system. For example doubling CO_2 from its pre-industrial level would increase the net flow of energy to the surface by 4 W/m^2. When there is a positive imbalance in radiative forcing, the Earth's surface must heat up until energy escaping to space balances incoming energy from the Sun.

Rebound effect The behavioral response to improvements in efficiency by which people increase their use of the more efficient activity. For example, if cars double their fuel economy but the price of gasoline is unchanged, the cost of driving a given distance drops by half so people may drive more. This increase in driving partly cancels the reductions in gasoline use achieved by making the cars more efficient. Estimates of the size of rebound effects vary widely, but they are unlikely to offset more than 10 to 20 percent of the initial savings.

Renewable energy sources Energy resources that draw on a continuously available flow of natural energy input, rather than drawing down a fixed stock of stored energy. Major renewable energy sources include solar, wind, hydroelectric, and biomass energy.

Scenario A description of potential future conditions produced to inform decision-making under uncertainty. In analyses of climate change, scenarios of future trends in greenhouse-gas emissions are used to produce climate-model projections of potential future climate change, and to analyze mitigation goals and their costs.

Scientific method The process of scientific investigation, involving the three-part logical structure of making an informed guess (a "hypothesis") about how the world works, identifying what the hypothesis implies for observable evidence, then testing the hypothesis by looking at the evidence.

Sectoral regulation Regulation of greenhouse-gas emissions that targets specific business sectors, products, or technologies, rather than the entire economy. Examples would include regulation of the fuel economy of vehicles, of the emissions intensity of steel production, or of the energy performance of buildings.

Solar variability Variations in the energy output of the Sun. The Sun's output varies slightly on time-scales of years, decades, and longer. Because the Sun is the power source for the Earth's climate, such variations can cause climate change. Measurements of solar output over the past few decades, however, have ruled out solar variability as the cause for any but a small fraction of the recent rapid warming.

Upstream regulation Regulation of greenhouse-gas emissions applied at the point where a unit of fossil fuel is extracted, imported, or processed, by applying an emission tax or requiring a permit at that point. The cost of the tax or permit becomes embedded in the price of the fuel as it moves through the economy, raising the cost of goods and services that use carbon-based energy. Upstream regulation is distinguished from regulating downstream, at the point where the fuel is burnted and the emissions occur. The advantage of upstream regulation is that it can cover most fossil fuel used in the economy by regulating at a small number of points – e.g., mines, refineries, pipelines, and ports – rather than at millions of homes, vehicles, and offices where actual emissions occur.

Urban heat island effect A warming of temperatures in urban built environments relative to nearby rural areas, caused principally by roads and buildings tending to be darker and absorb more sunlight than natural land cover. This effect can introduce a spurious warming trend into temperatures as cities develop and expand into formerly rural areas, which is removed by statistical methods in calculating long-term temperature trends.

World Trade Organization (WTO) The international organization that oversees negotiations and rules concerning international trade to promote free trade and limit discriminatory trade practices. Proposed trade measures to motivate and enforce greenhouse-gas reductions, such as import taxes levied on imports in proportion to the emissions generated in their production, would have to comply with WTO rules for non-discriminatory trade.

Bibliography

ACIA (2005). *Impacts of a Warming Arctic: Arctic Climate Impact Assessment*. Cambridge, UK: Cambridge University Press. http://www.acia.uaf.edu

Aldy, J. E., Krupnick, A. J., Newell, R. G., Parry, I. W. H., and Pizer, W. A. (2009). *Designing Climate Mitigation Policy*, Discussion paper 08–16 (May). Washington, DC: Resources for the Future. http://www.rff.org/RFF/Documents/RFF-DP-08-16.pdf

Aldy, J. E. and Stavins, R. N., eds. (2009). *Post-Kyoto International Climate Policy: Summary for Policymakers*. Cambridge, UK: Cambridge University Press.

Archer, D. (2007). *Global Warming: UnderStanding the Forecast*. Malden, MA: Blackwell Publishing.

CCSP (US Climate Change Science Program) (2006). *Temperature Trends in the Lower Atmosphere: Steps for Understanding and Reconciling Differences*. Karl, T. R., Hassol, S. J., Miller, C. D., and Murray W. L., eds. Washington, DC: Climate Change Science Program. http://www.climatescience.gov/Library/sap/sap1-1/default.php

(2009). *Global Climate Change Impacts in the United States. Unified Synthesis Product*. Karl, T. R., Melillo, J. M., and Peterson, T. C., eds. Washington, DC: Climate Change Science Program. http://www.climatescience.gov/Library/sap/usp/default.php

Clarke, L., Edmonds, J., Jacoby, H., Pitcher, H., Reilly, J., and Richels, R. (2007). *Scenarios of Greenhouse Gas Emissions and Atmospheric Concentrations*. Synthesis and Assessment Product 2.1a. Washington, DC: US Climate Change Science Program. http://www.climatescience.gov/Library/sap/sap2-1/default.php

Council on Foreign Relations (2008). *Confronting Climate Change: A Strategy for U.S. Foreign Policy*. Independent Task Force Report No. 61, G. E. Pataki and T. J. Vilsack, chairs. New York: Council on Foreign Relations.

Dessler, A. E. (2000). *The Chemistry and Physics of Stratospheric Ozone*. San Diego, Academic Press.

Emanuel, K. (2007). *What We Know About Climate Change*. Cambridge, MA: MIT Press.

Farrell, A. and Jäger, J., eds. (2005). *Assessments of Regional and Global Environmental Risks: Designing Processes for the Effective Use of Science in Decision-making*. Washington, DC: Resources for the Future.

IPCC (2001). *Climate Change 2001: The Scientific Basis*. Contribution of Working Group I to the Third Assessment Report of the Intergovernmental Panel on Climate Change. Houghton,

J. T., Ding, Y., Griggs, D. J., Noguer, M., van der Linden, P. J., Dai, X., Maskell, K., and Johnson, C. A., eds. Cambridge, UK: Cambridge University Press.

(2000). *Emission Scenarios.* Special report of the Intergovernmental Panel on Climate Change. Nakicenovic, N. and Swart, R., eds. Cambridge, UK: Cambridge University Press.

(2007a). *Climate Change 2007: The Physical Science Basis.* Contribution of Working Group I to the Fourth Assessment Report of the Intergovernmental Panel on Climate Change. Solomon, S., Qin, D., Manning, M., Chen, Z., Marquis, M., Averyt, K.B., Tignor, M., and Miller, H.L., eds. Cambridge, UK: Cambridge University Press.

(2007b). *Climate Change 2007: Impacts, Adaptation and Vulnerability.* Contribution of Working Group II to the Fourth Assessment Report of the Intergovernmental Panel on Climate Change. Parry, M.L., Canziani, O.F., Palutikof, J.P., van der Linden, P.J., and Hanson, C.E., eds. Cambridge, UK: Cambridge University Press.

(2007c). *Climate Change 2007: Mitigation of Climate Change.* Contribution of Working Group III to the Fourth Assessment Report of the Intergovernmental Panel on Climate Change. Metz, B., Davidson, O., Bosch, P., Dave, R., and Meyer, L., eds. Cambridge, UK: Cambridge University Press.

(2007d). *Climate Change 2007: Synthesis Report.* Core Writing Team, Pachauri, R.K, and Reisinger, A., eds. Cambridge, UK: Cambridge University Press.

Jasanoff, S. (1990). *The Fifth Branch: Science Advisors as Policymakers.* Cambridge, MA: Harvard University Press.

Kuhn, T. (1962). *The Structure of Scientific Revolutions.* Chicago: University of Chicago Press.

Lempert, T. (2009). *Setting Appropriate Goals: A Long-term Climate Decision.* Workshop paper, "Shaping Tomorrow Today: Near-term steps towards long-term goals." Santa Monica: RAND Pardee Center. http://www.rand.org/international_programs/pardee

Lisiecki, L. E. and Raymo, M. E. (2005). A Pliocene-Pleistocene stack of 57 globally distributed benthic delta O-18 records, *Paleoceanography*, **20**, PA1003, DOI: 10.1029/2004PA001071.

Mitchell, R., Clark, W.C., Cash, D., and Dickson, N., eds. (2006). *Global Environmental Assessments: Information and Influence.* Cambridge, MA: MIT Press.

Moss, R., Edmonds, J., Hibbard, K., Manning, M., Carter, T., Emori, S., Kainuma, M., Kram, K., Manning, M., Meehl, J., Mitchell, J., Nakicenovic, N., Riahi, K., Rose, S., Smith, S., Stouffer, R., Thomson, A., van Vuuren, D., Weyant, J., and Willbanks, T. (2009). Representative concentration pathways: a new approach to scenario development for the IPCC Fifth Assessment Report, *Nature* (in press). www.pnl.gov/gtsp/publications/2008/papers/20080903_nature_rcp_new_scenarios.pdf.

Nordhaus, W. (2008). *A Question of Balance: Weighing the Options on Global Warming Policies.* New Haven: Yale University Press.

Parson, E. A. (2003). *Protecting the Ozone Layer: Science and Strategy.* New York: Oxford University Press.

(2008). *The Long Haul: Managing the Energy Transition to Limit Climate Change.* Workshop report, August 2008. www-personal.umich.edu/~parson/website/research.html

Petit, J. R., Jouzel, J., Raynaud, D., Barkov, N. I., Barnola, J.-M., Basile, I., Bender, M., Chappellaz, J., Davis, M., Delaygue, G., Delmotte, M., Kotlyakov, V. M., Legrand, M., Lipenkov, V. Y., Lorius, C., Pépin, L., Ritz, C., Saltzman E., and Stievenard, M. (1999). Climate and atmospheric history of the past 420,000 years from the Vostok ice core, Antarctica, *Nature*, **399**, 429–436.

Ruckelshaus, W. D. (1985). Risk, science, and democracy, *Issues in Science and Technology*. **1**:3, Spring, 19–38.

Smith, J. B., Schneider, S., Oppenheimer, M., Yohe, G., Hare, W., Mastrandrea, M., Patwardhan, A., Burton, I., Corfee-Morlot, J., Magadza, C., Füssel, H.-M., Pittock, A., Rahman, A., Suarez, A., and van Ypersele, J.-P. (2009). Assessing dangerous climate change through an update of the IPCC 'reasons for concern', *Proceedings of the National Academy of Sciences*, **106**, 4133–4137.

Stern, N. H. (2007). *The Economics of Climate Change: The Stern Review*. Cambridge, UK: Cambridge University Press.

Tester, J. W., Drake, E.M., Driscoll, M.J., Golay, M.W., and Paters, W. A. (2005). *Sustainable Energy: Choosing Among Options*. Cambridge, MA: MIT Press.

US National Research Council (2006). *Surface Temperature Reconstructions for the Last 2,000 Years.* Board on Atmospheric Sciences and Climate. Washington, DC: National Academies Press.

Weart, S. R. (2003). *The Discovery of Global Warming*. Cambridge, MA: Harvard University Press.

Zachos, J., Pagani, M., Sloan, L., Thomas, E., and Billups, K. (2001). Trends, rhythms, and aberrations in global climate 65 Ma to present, *Science*, **292**, 686–693.

Index